The

EMOTIONAL NATURE *of* QUALITATIVE RESEARCH

INNOVATIONS
in PSYCHOLOGY

Series Editor
Charles R. Figley, Ph.D.
Florida State University

Editorial Board

PUBLISHED TITLES

Innovations in
Psychology

The

EMOTIONAL NATURE *of* QUALITATIVE RESEARCH

Edited by
Kathleen R. Gilbert, Ph.D.

CRC Press
Boca Raton   London   New York   Washington, D.C.

Library of Congress Cataloging-in-Publication Data

Catalog record is available from the Library of Congress.

From the series editor

Emotional intelligence, a concept coined by Goleman,[1] is the ability to discern, utilize, and shape the emotional reactions of self and others. Emotions represent an area rich in tradition and span fields of study, as well as culture and time.

This is the latest book in the series, *Innovations in Psychology*. Professor Kathleen Gilbert, through this book, introduces the field to an important new area, emotion-focused research.

Emotions include, but are not limited to, stress and flow or lack of it (the manifestation of emotions), feelings (that represent and cause the emotion stress), and emotional controls (our efforts to modulate our emotional effects). Emotions account for an extraordinary amount of the human experience. It is remarkable how little we know about this phenomenon. It is about time that we address the emotional nature of research.

"How do we manage the emotions stimulated by researchers? What does it mean if the researcher is profoundly affected by the experience, even to the extent of going through a spiritual epiphany?" Dr. Gilbert, a professor at Indiana University for many years, addresses these and other related questions in a book for researchers and practitioners who wish to apply qualitative (as opposed to simply quantitative/data analysis-driven) research methods to human research.

The tradition of this book series represents substantial and trendsetting innovations in psychology, has set new agendas for burnout in families,[2] and has opened new areas of clinical innovation by introducing energy psychology,[3] traumatology,[4,5] and poetry therapy.[6]

In a period of substantial scholarly activity in the social and medical sciences, giving voice to research participants is suddenly more acceptable beyond anthropology and those practitioners of qualitative research methods. This trend is part of a larger cultural trend, even a global trend.

In part this is due to the impersonal aspects of the internet, the growing concentration of people in the cities, and other demographic explanations. However, some suggest that psychology has lost its heart and soul. Gone are the influences of social psychology that led to major breakthroughs in race relations, human rights, war prevention, and other important social causes. Goleman's[1] book on emotional intelligence highlighted this important area of inquiry far beyond psychology. Similarly, Joseph LeDeux's[7] classic application of cognitive neuroscience to understanding human emotion is an important indicator of this trend.

With this book, psychologists have the opportunity to move toward greater understanding of themselves as researchers. The emotional components of studying research participants often get lost in reports of quantitative data only. We need the qualitative picture for our frame of reference.

As a professor since 1971, I have seen the academy reward quantitative researchers and perceive as suspect any effort to incorporate subjectivity. As an editor for three scholarly journals and a reviewer for dozens of others, qualitative research has been an enigma. No longer.

More than in any other period of history there is great interest in the research participants' emotional voices offered through interviews, oral histories, content analysis, naturalistic analysis, and the various methods of ethnographers.

This international collection is a contribution to any library devoted to methods of studying human beings. Timeless and universal, this collection of personal discussions of emotion in the research enterprise provokes the reader to think. Students of emotion will find it useful and a reflection of their own experiences as scholars in this area. Researchers will find it intriguing because it challenges conventional methodology and assumptions. This is why it is in this book series, *Innovations in Psychology*.

Are you aware of an important innovation in psychology? The Editorial Board and I would like to know about it. Nominations for books should be sent to me via e-mail at *cfigley@mailer.fsu.edu*. Comments on this and other books in the series are especially welcome.

Charles R. Figley, Ph.D.
Series Editor

References

1. Goleman, D., *Emotional Intelligence: Why It Can Matter More than IQ for Character, Health and Lifelong Achievement*. New York: Bantam Books, 1995.
2. Figley, C. R. (Ed.), *Burnout in Families: The Systemic Costs of Caring*. Boca Raton: CRC Press, 1998.
3. Gallo, F., *Energy Psychology. Explorations at the Interface of Energy, Cognition, Behavior, and Health*. Boca Raton: CRC Press, 1999.
4. French, G. and Harris, C. J., *Traumatic Incident Reduction (TIR)*. Boca Raton: CRC Press, 1997.
5. Lipke, H., *EMDR and Psychotherapy Integration: Theoretical and Clinical Suggestions with Focus on Traumatic Stress*. Boca Raton: CRC Press, 2000.
6. Mazza, N., *Poetry Therapy: Interface of the Arts and Psychology*. Boca Raton: CRC Press, 1999.
7. LeDeux, J., *The Emotional Brain: The Mysterious Underpinnings of Emotional Life*. New York: Touchstone, 1998.

About the Editor

Kathleen R. Gilbert, Ph.D., is an Associate Professor of Applied Health Science at Indiana University, Bloomington, where her work focuses on family and health issues. She co-authored a qualitative study of loss, *Coping with Infant or Fetal Loss: The Couple's Healing Process*, and co-edited *Research and Theory in Family Science*. She also has presented and published numerous journal articles and book chapters on the dynamics of loss, grief, and trauma in a dyadic and family context. In addition, she developed *Grief in a Family Context*, a bi-level, graduate and undergraduate course that is delivered over the internet. Dr. Gilbert's research interests in "cybertechnology" include the use of the World Wide Web as a medium for memorializing a loved one, the internet as a tool for grief management and reconstruction, and electronic communities for disenfranchised grievers.

Contributors

Hilary Davis, Ph.D., recently completed her doctoral studies in Medical Sociology for the University of Sheffield, England. At the time this work was conducted she was a research fellow managing a variety of projects, including the evaluation of patient information technologies (leaflets, interactive video, etc.) in both primary and secondary healthcare settings. Her research interests include: the role of emotions in ethnographic research in organizations, the relationship between technology and patient care, and the evaluation of new information technologies in hospitals and other health care environments.

Jennifer Harris, Ph.D., is a Senior Research Fellow in the Social Policy Research Unit, University of York, England. Her first book, *The Cultural Meaning of Deafness* was published in 1995, and her second *Deafness and the Hearing* in 1997. Dr. Harris has, over many years, pursued her research interests in the fields of qualitative methodology and disability studies. Her recent publications include "Self Harm: Cutting the Bad Out of Me," a study undertaken by correspondence with self-harming women, in the *Qualitative Health Research Journal*.

Annie Huntington, Ph.D., is a Lecturer at the University of Salford, England. Dr. Huntington has worked as a nurse, social worker, teacher and psychotherapist within health and social care over the last 20 years. Her recent publications include a social work monograph for the University of East Anglia, *Social Work With Children and their Families: Professional Dilemmas*, as well as several articles in social work journals including "The Reflexive Social Work Researcher" for *Social Work and Social Sciences Review*.

Christine E. Kiesinger, Ph.D., is Assistant Professor and Chair of the Department of Communication at Southwestern University in Georgetown, Texas. Dr. Kiesinger's scholarship is centered in the area of identity with a specific focus on the lives and relationships of anorexic and bulimic women. Mostly biographical and autobiographical in nature, Dr. Kiesinger's research aims to capture the texture of emotional experience. Dr. Kiesinger's work is often cast in narrative and poetic form and is written to evoke connection and to inspire healing.

Jim Mienczakowski, Ph.D., devised a construct of critical ethnodrama (health research which influences practical change in health settings) for his doctoral studies which has evolved into a new and important form of ethnographic practice for arts and education practitioners and is attracting increasing international recognition. Dr. Mienczakowski's current research focuses on the emotional trauma of cosmetic surgery, trajectories of recovery from sexual assault and includes submissions to state government social impact reports affecting community change. As an invited expert witness, he provided testimony for the 1999 HCCC report into cosmetic surgery. He currently serves on an advisory group which assesses health promotional theater involving issues of youth suicide. He is Foundation Dean of the Faculty of Education and Creative Arts, CQU and Dean of the Central Queensland Conservatorium of Music.

Stephen Morgan, M.Ed., is a Registered Psychiatric Nurse at the Faculty of Nursing and Health, Griffith University, Australia. In addition to developing innovative research frameworks in performance and representation, He currently lectures in counseling skills, communication, mental health and psychiatric assessment with the Faculty of Nursing and Health at Griffith University. He is also conducting research into school suicide with the Australian Institute for Suicide Research and Prevention and is currently a member of the Queensland Health Mental Health Morbidity Committee. He has also completed a period of consultancy work with Lifeline Australia.

Paul C. Rosenblatt, Ph.D., is Professor of Family Social Science at the University of Minnesota. He has carried out many qualitative research projects. Among the book-length reports of his qualitative research are *Parent Grief: Narratives of Loss and Relationship*; *Multiracial Couples: Black and White Voices*, *Farming Is in Our Blood: Farm Families in Economic Crisis*; *The Family in Business*; and *Bitter, Bitter Tears: Nineteenth Century Diarists and Twentieth Century Grief Theories*. His current projects include research focused on grieving families, multiracial/multicultural families, and Korean families. He also is exploring the use of fiction to present qualitative research findings.

Lynn Smith, B.C.A., is a Former Secondary School Art and Social Education Teacher who holds Master Teacher level 1 accreditation. During her professional teaching career, and concurrent with her teaching practice, she was an active member of various professional teaching organizations including the Northern Territory Board of Studies (NTBOS); the Aboriginal Education Standing Committee of the NTBOS; the Art Educators of the NT; the Australian Teaching Council Board; and an executive member and Women's Officer for the AEU/NT. She recently began a Master of Education program examining the impact of her theory of "ethnovisualisation" on society. Lately, her professional career has altered course and she is now employed as the Research Associate to the Dean of the Faculty of Education and Creative Arts at Central Queensland University.

Cynthia M. Stuhlmiller, DNSc., is currently Professor of Mental Health Nursing at The Flinders University of South Australia. She was the Foundation Chair of Mental Health Nursing at University of Technology, Sydney from 1997–2000. Previously, she held academic appointments in Norway, New Zealand, and the U.S.A. Her clinical background includes 20 years in a variety of community and inpatient mental health settings. She has worked extensively with people exposed to traumatic stress from war, disaster, and other extreme conditions. Her research interests include responses to disaster, rescue and emergency work, seasonal variation, narrative picturing, and the dangers of diagnostic disordering.

Lisa M. Tillmann-Healy, Ph.D., is an Assistant Professor of Communication at Rollins College in Winter Park, Florida. She has taught courses in the areas of interpersonal and family communication and media studies. Her most recent scholarship has centered on close relationships and qualitative methods. She is publishing a book based on her dissertation fieldwork, an ethnographic study of a community of gay men that explores the possibilities for friendship across sexual identity.

Emma Wincup, Ph.D., is a Lecturer in Sociology at the University of Kent at Canterbury, England. Her main research interests lie in the sociology of crime and deviance and include gender, crime, and criminal justice and substance misuse. She has published widely on a range of methodological issues including *Qualitative Research in Criminology,* and *Doing Research on Crime and Justice.*

Contents

Acknowledgments

This book would not be possible without the efforts of many people who shared my vision and understanding.

My development as a researcher and scholar began with my doctoral research project, supervised by Dr. Charles R. Figley who has continued to mentor me through the years, and I am most grateful for that. He facilitated the publication of this book and inspired me as I worked to bring it to completion. I am particularly appreciative of his assistance.

Colleagues, both here at Indiana University and around the globe have been a source of support and encouragement. Conversations with my colleague, Dr. Deborah Fravel, here at Indiana University, helped me focus my thinking as I wrote my chapters. Dr. Karen Schmid, who shared my early enthusiasm for this topic, has been an immense help, as were the many conference attendees at our brainstorming session at the 1994 Annual Conference of the National Council on Family Relations.

The 1996 Fourth International Conference on Social Science Methodology, held at the University of Essex in Colchester, England, allowed me to meet more than half of the authors, and I would like to thank the program coordinators for that opportunity.

Many thanks go to the authors of the chapters who represent three countries, bringing cultural variety and enrichment to this book. In many instances, these authors took professional risks to present their personal accounts so clearly. They also had to deal with e-mail after e-mail of requests for modifications, but they responded with wonderfully crafted chapters.

I would like to thank Dr. Tony Mobley, Dean of the School of Health, Physical Education and Recreation at Indiana University, my former Department Chairperson, Dr. James Crowe, and my current Chairperson, Dr. Mohammad Torabi, for the support they have shown over the years for my "atypical" approach to research. The grant money, particularly from the departmental internal grants program, has facilitated my research, and indirectly moved me forward in my understanding of the emotional nature of qualitative research.

My relationship with Barbara Mercer and Joelene Bergonzi, Indiana University graduate students, was a source of great pleasure throughout the book development process. Their careful reading of the chapters, from the perspective of students new to qualitative research, was immeasurably helpful to the process, and enriched my understanding of the intensity of the first research experience for someone in the vulnerable position of a student.

Thanks to my secretary, Donna Johnson, for her encouragement and support and for clerical assistance she provided on the project.

Finally, my family has been integral to my completion of this volume. My husband, Dr. Steven W. Gilbert and my daughters, Kimberly Ellerthorpe and Rebecca Gilbert, provided a stabilizing force, even as I alternated between efforts on the book and the many outfits I concurrently sewed for Kim's wedding to Shawn.

Dedicated to my family

Steven Wilson Gilbert

Kimberly Gilbert Ellerthorpe

Rebecca Gilbert

For the support and inspiration they have provided to me

section one

A detailed look

chapter one

Introduction: why are we interested in emotions?

Kathleen R. Gilbert

Contents

Introduction

It seems that only a short time ago qualitative research methods were viewed within traditional fields like psychology and sociology with suspicion and, in some cases, outright contempt. The intensely personal involvement of the researcher in the data collection and analysis process was seen as questionable, lacking rigor, and as anecdotal "talky-talk." Although an examination of scholarly journals shows that this view persists, many have come to see the unique value of qualitative methods for the exploration of social and psychological phenomena. An upsurge of interest in qualitative methods is seen in various branches of psychology, most notably counseling, social, and narrative psychology. Nevertheless, neophyte researchers as well as more seasoned quantitative researchers new to these methods may bring unexpected "baggage" to their efforts as qualitative researchers. This baggage, if unexamined, can adversely influence the research process.

0-8493-2075-5/01/$0.00+$.50
© 2001 by CRC Press LLC

My experiences as a teacher of qualitative methods, dissertation supervisor, and professional presenter and writer have taught me that students, in their naturally ambiguous state, are especially interested in learning how to deal with the emotional content of their research efforts. Having been indoctrinated with the philosophy of managing, rather than integrating, emotions into the research process, I can certainly relate.

More recently, I have begun to hear the same questions from faculty originally trained as quantitative researchers who are now working in this realm or supervising qualitative dissertations. They are surprised and even startled by the emotions generated by their research efforts. They ask: how do we manage these emotions? Should we even think in terms of managing emotions? What is the proper balance of allowing and controlling emotions? Are emotions appropriate? Should they be used as a part of the research process? Should they be hidden and kept from others? What does it mean if the researcher is profoundly affected by the experience, even to the extent of going through a spiritual epiphany?

This volume addresses the personal and emotional nature of qualitative research which involves a process of entering the subjective world of research participants. In order to do this, qualitative researchers must "become the research instrument," a phrase commonly used in reference to engagement in the qualitative research process. In doing this, they naturally draw on elements of their own subjectivity. Although the tendency among researchers has been to mirror the expectation of objectivity maintained by quantitative research,[1] increasingly researchers are expected to be emotionally affected by the research process.[2] Although this is true, we evidence ambiguity toward acknowledging publicly this effect. "Classic ethnographies either omit a researcher's emotions or relegate them to a preface or appendix."[3] Emotions may also be presented as separate "confessional tales" like Katz-Rothman's[4] personal account of her response to the experience of women who underwent amniocentesis and, in many cases, decided to terminate a much-wanted pregnancy. Thus, even with our increasing comfort with emotions in qualitative research, we continue to struggle with how deeply to integrate our emotions into the research process.

A bit of history

The genesis of this volume, and a major transitional point in my own interest in the relationship between emotions and the qualitative research process, was a single comment in a five-minute conversation. In the early 1990s, I had a brief conversation with Murray Strauss, one of the two "fathers" of grounded theory,[5,6] about what I thought was a unique personalization of the subject matter of my dissertation research project. In our conversation, I described a series of dreams that I had experienced while working on my dissertation project, which looked at grief and coping among parents who had lost a child in pregnancy or infancy.[7,8] These dreams clearly were related to the dissertation focus and had a clear chronology. The early dreams were pervaded by a vague sense that my daughters (at the beginning of my project, ages 9 and 11) were somehow threatened. Over time, the dreams became increasingly threatening concerning my daughters, particularly as I began to conduct interviews. They continued to be threatening until I experienced a dream in which the body of my older daughter was found. As I spoke with Strauss, I could remember (and, as I write this, still remember) the intensity of my emotional response to this dream. Even now, I can visualize the position of her body, and the fact that we were in an alley and that the building behind her was old and constructed of chipped Chicago brick. I recall that the surface of the alley was damp, the light was penumbral, and the colors were muted. As I reviewed the dream over and over (sometimes willingly, sometimes not) and continued to interview the bereaved parents, a curious thing began to happen. The dreams progressed to being endangering (but not deadly), clearly threatening, and then vaguely threatening, until I had a final dream in which I actually dreamt that I was dreaming and told myself to stop.

Strauss's response to this was "That sounds like a sophisticated form of going native." He then went on to advise me on ways in which I might be able to manage my emotional response. I must confess that I heard little of his advice as I was still ruminating on (and smarting from) his initial comment. The idea of "going native"[9] had always been presented to me as a negative, something that would remove me from the role of scientist and shift me into the role of someone who had become too close to the phenomenon under study — something to be avoided at all costs. At the same time, I felt conflicted because I had experienced the dreams as a positive, albeit painful, side effect of my interviews. I consistently felt that I had a better sense of what the parents were trying to communicate to me when I had experienced one of these dreams prior to conducting an interview. I also believed that they afforded me a safe way to deal with my own internal parental terrors that were being triggered by these interviews with bereaved parents[10] who are good people unable to prevent the death of their much-loved child. How was I to relate to them in a meaningful way and not be overwhelmed by their pain and my fears for my daughters? These caring people seemed so much like me and my husband. Should I simply focus on managing my emotions in order to retain some form of objectivity? Could I legitimately hold myself apart from the personal and emotional? I came to believe that it is not possible and that it might not be as desirable as I had once thought.

My interest in the idea of using one's own emotions in the research process continued to grow, although I found myself frustrated by how little had been written on positive aspects of bringing our emotions into the research process and how hard I had to search to find that. Although I found some excellent resources from other research traditions,[2,11-13] I found none in the psychology literature that was my primary focus at that time.

In 1993, I co-led a preconference brainstorming session with Karen Schmid[14] on emotions in qualitative research at the annual conference of the

National Council on Family Relations (NCFR). This session was later summarized in Gilbert and Schmid.[15] Reflecting where I was in my progression, that summary focuses on the discussion of managing and balancing forces and their threat to professional identity. The discussion of the positive aspects of emotions in the research process was minimized. As Kleinman and Copp[3] conjectured "Perhaps we also fear that quantitative sociologists will find out the truth — that our work is as subjective as they suspected, and we fear that our work (and we) will be rejected as inadequate."

Even with my interest in the emotional involvement of the researcher in the research process, as I conducted this session, I described myself as an "information conduit" and saw myself as a means by which the stories of the participants could be conveyed. Although I felt emotionally tied to parents who had lost a child, I believed that I contained my emotions, particularly in the writing phase. It was *their* stories that I conveyed. My own involvement, I was sure, was that of organizer and analyst. I presented myself as having kept my Self out of the process. I was disabused of this notion by members of the group, particularly the feminist researchers. Their stance was that, at minimum, researchers as well as participants filter everything they experience in the research process through their own biases, views, and feelings. This concept of a "co-authored narrative," in which researcher and researched contribute to a final text that moves through several filters from the initial telling through the final interpretation and reporting,[13] ends in a collaboration of sorts with the reader of the text. It is addressed by several authors in this volume.

I found myself with more questions than answers after this brainstorming session, but with the clear awareness that I was not alone in my response to the research process. In July 1996, I had the opportunity to propose and organize a conference stream for the Fourth International Conference on Social Science Methodology, held at the University of Essex in Colchester, England. Four conference sessions were dedicated to the theme "The Emotional Nature of Qualitative Research," and several of this volume's chapter authors participated in these sessions. In addition, at this international conference, input was solicited from audience members on what they would include if they could design a volume on this subject. Based on the conversation with this group and later conversations with others interested in qualitative research, along with my own ruminations, ideas for topics were generated and manuscripts on these topics were solicited. The final product is this volume.

The design of this volume

This volume addresses the personal and emotional nature of qualitative research and the place and purpose of emotions in the research process. It discusses the impact on others involved in the research process, as well as the researcher, and extends beyond the relationship between the researcher and researched to the researcher's roles of study administrator, teacher, friend, and family member. In each chapter, authors were asked to write about some aspect of what they saw to be the emotional nature of qualitative research, without defining the term "emotions" for them. The authors were allowed to approach the relationship with the greatest latitude.

Much emphasis has been placed on the dangers of emotions and the need for caution throughout the research process. Here, we take the view that, although this may be true, emotions also contribute positively to the process. The authors also address this aspect of the relationship between qualitative research and emotions.

The authors of this volume were selected intentionally from a range of levels of experience. The first exposure to qualitative research may have the greatest impact (it did for me, and I have been told by others that this was the case for them); therefore, Chapters 2 through 5 (Section 1) were written by authors who are new to qualitative research. Meanwhile, the ongoing exposure to the emotional nature of qualitative research can have a transformative impact; therefore, Chapters 6 through 10 (Section 2) are written by more seasoned authors who have been doing qualitative research for several years. The intent is to draw on the wealth of experiential knowledge of the practiced researchers as well as the discovery process of those new to the process. One unique aspect of this volume is that the text is multinational, with authors representing the U.K., the U.S., and Australia. Interestingly, even as the perspectives are influenced by the culture within which each author lives, overarching themes emerge from the works of various authors.

Section 1 is made up of chapters written by researchers at the beginning of their research careers. Each chapter involves a doctoral dissertation, and collectively they document the value of research training and supervision that is informed in the emotional nature of qualitative research. They also document the drawbacks of entering the qualitative research arena with training and supervision that neglect this aspect of qualitative research.

In Chapter 2, Emma Wincup recounts her experiences while interviewing women awaiting trial. She discusses how, even in the feminist research tradition which has promoted use of the qualitative paradigm, discussion of the emotional nature of the research process is frowned upon. Wincup describes her own experience in the field and ways in which she used her emotional response to what she encountered during the fieldwork for her study.

Chapter 3 by Hilary Davis demonstrates that it is not necessary for the focus of the study to be on an emotional topic for the research experience to be an emotional one. Unlike Wincup, who had been encouraged to study something she felt angry about, one might have anticipated that Davis's study of the use of information and other technology in a hospital setting would have been "emotion-free." In a richly detailed account, she clearly documents how the exposure to the culture of the hospital, in a poorly defined, socially ambiguous role (requiring the hospital staff to define her identity, sometimes in very negative and complicating ways), led to intense emotion. Triggers of emotion abounded: sights, sounds, smells, and events in which hospital staff seemed to engage in callous and uncaring behavior. Ultimately, she argues that ethnographers recognize the interplay between the emotional and intellectual work and that, even in the case of topics that are considered unemotional, an emotional dimension exists.

Cynthia M. Stuhlmiller, in Chapter 4, uses her experiences in studying the response of disaster workers during the 1989 Oakland, California earthquake to discuss therapeutic transformation resulting from participation in a narrative research study. She was a part of a therapeutic intervention team, and yet it was through narration of their stories for a research study that participants experienced a therapeutic effect, not through clinical debriefing. That is, participants were changed, in positive ways, by telling their stories to a supportive and caring person in a format that was nonthreatening — something that would not have been the case for these individuals had this been a clinical interview.

Chapter 5, by Lisa M. Tillmann-Healy and Christine E. Kiesinger, chronicles their interrelated and interwoven studies of women dealing with bulimia. The authors follow the chronology of their co-narrated piece, in which they interviewed each other and grew to understand themselves as well as each other through the process of their studies. As they move back and forth to tell their story, the title of the chapter, "Mirrors", becomes crystal clear.

Section 2 takes a longer view across studies and across time and life experience. The chapters in this section demonstrate personal development and maturation in the authors' progression as researchers. These chapters show qualitative research to be an evolutionary emotional experience. We come to have a different appreciation of the emotional nature of qualitative research and with each exposure we learn more.

Chapter 6, by Paul C. Rosenblatt, is a very personal exploration of his own spiritual transformation that resulted from the personal and emotional character of his qualitative research. Beginning as a quantitative researcher, he went through changes in how he viewed himself and the world. With this came emotional highs and lows, insights, and some effects that seemed negative and others quite positive.

Jennifer Harris and Annie Huntington, in Chapter 7, look at the potential psychotherapeutic effects of qualitative research. As did Wincup, these authors also use a feminist perspective, arguing that it is "important to focus on our emotional and cognitive responses when in the role of researcher." They also discuss, in detail, various ways of understanding how the term "emotion" is used, and the difficulty of using a single definition. Finally, they discuss what we might learn from psychotherapeutic practice in the conduct of qualitative interviews and some of the cautions we should keep in mind.

In Chapter 8, I look at the impact of the emotional nature of qualitative research on an unanticipated population — staff and students involved in data management. I look at the effect of being exposed to and dealing with fragmentary "bits and pieces" of people's lives, and suggest how we might, as research directors, manage to reduce the negative effects of this exposure.

Ethnodrama and its potential use of emotion to make social change is the focus of Chapter 9, by Stephen Morgan, Jim Mienczakowski, and Lynn Smith. In this chapter, the authors describe how they use interview transcripts to develop play scripts, how these are performed, and the intended and unintended effects on participants in the research, the writing and performing teams, as well as on audiences. Emotions are consciously used here, as opposed to the efforts described by other authors in this volume. Included in Chapter 9 is a list of cautions about using this technique, based on both positive and negative effects the authors experienced in their own work.

Chapter 10, written by Jim Mienczakowski, Stephen Morgan, and Lynn Smith, extends Chapter 9 by discussing the effects of studying "night workers," that is, individuals who work in "occupations, professions, and activities which are both acknowledged as essential to social existence but are marginalized in the everyday experience, thoughts, and recognitions of most people." On a personal note, and because I research grief and loss, I have experienced the sense of being "pushed into the shadows" when I describe my research interests. People have asked me questions such as "How can you study that? That is so morbid!" followed by a rapid change of topic on the person's part or a sidling or turning away from me as they seek to "avoid the specter of death."

Our interest here is in sensitizing the reader to the fact that the research process is emotional and that can be a positive as well as a negative. By using emotions intelligently, we can be better researchers and our research output can be more truthful. This volume is intended to present the reader with a series of personal accounts of qualitative researchers who have performed and continue to perform their "balancing acts" with regard to closeness to the phenomenon under study, and with both negative and positive aspects of the emotional nature of qualitative research.

What we mean by "emotions"

Some qualitative researchers believe that the ultimate goal of research is to enter the world of others in such a way as to allow the researcher to see life through their eyes. In order to do this, one cannot see this as a purely intellectual exercise, but as a process of exploration and discovery that is felt deeply — that is, research is experienced both intellectually and emotionally. But what do we mean when we speak of emotions? I found, in my initial exploration of the concept, a great deal of ambiguity and ambivalence toward emotions starting with a lack of a clear definition. Some terms that other qualitative researchers shared with me as I began to study this process include feelings, sensations, drives; the personal; that which is intimate; personally meaningful, possibly overwhelming; being touched at a deeper level; something that comes from somewhere within ourselves; and that which makes us truly human.

These comments reflect an essential problem faced in efforts to incorporate the emotional into the research process — the problem of a lack of clarity in defining what we mean by the term "emotions." Various authors have pointed this out including Jagger who noted the "variety, complexity, and even inconsistency of the ways in which emotions are viewed, both in daily life and in scientific contexts."[16] The range of phenomena covered using this term is immense, as are the criteria for choosing one definition over another.[16,17] The result, according to Sarbin, is that "(t)he word 'emotions' has drifted into opacity at the hands of physiologists, psychologists and phenomenologists. The imagery is diffuse and not easily communicated."[18] In Chapter 7, Harris and Huntington go into great detail on ways of approaching the concept of emotions in research and in common usage.

In this volume, we approach emotions in a variety of ways. They are varied and complex, unintentional and intentional, socially constructed, something to be actively engaged in, and interwoven with one's value system. We recognize that what we experience as emotion is basically physiological in origin; human reflexivity transforms these responses into what we define and experience as "emotions".[19]

Emotions are culturally defined and socially constrained.[19] They are more than physiological sensations, but are often experienced in this way. They guide our interpretations of what we experience and are shaped by our life experience. Many authors in this volume feel that they have been changed forever by the impact of their research on them. They have also seen similar effects that mirror their own on their informants and on the audiences of their finished texts.

How emotions impact the research process

As Jagger[16] notes, "Western epistemology has tended to view emotion with suspicion and even hostility." This is evidenced in the way it is (and is not) incorporated into research and the general lack of interest in public discussion of our emotional reactions. Emotions are rejected, because reason rather than emotion has been regarded as indispensable for acquiring knowledge, and emotion puts this at risk.[18] This can be seen in the following quotes: "The emphasis for years in psychology has been on cognition and rationality, and on ways of diminishing the influences of subjectivity and emotion in decision making and behavior."[17] "Mostly, sociologists have considered emotionality as a problem to overcome, something to avoid in order to do good research."[1] Researchers have been encouraged to control or even suppress their emotions, yet this "removal" of emotions from the research process does not mean that emotions are not present nor does it guarantee that the hidden emotions do not affect the research process.[16]

Ellis[1] has argued for the integration of emotional awareness into the research process by saying "Emotional sociology — consciously and reflectively feeling for ourselves, our subjects, and our topics of study, and evoking those feelings in our readers — is necessary for apprehending important aspects of social life, in particular, the lived experiences of emotions." The influence of emotion extends on all aspects of cognition and behavior,[17] and this extends to what we do in the research process.[1,16] Goleman's[19a] discussion of emotional intelligence, that is the ability to know and manage one's own emotions, to recognize emotions in others, and to handle relationships (i.e., handle emotions in others), has clear implications for high-quality research efforts. It is not the avoidance of emotions that necessarily provides for high quality research. Rather, it is an awareness and intelligent use of our emotions that benefits the research process.

If the qualitative researcher is to be the research instrument, then he or she must be fully aware of the nature of that instrument. What is at issue, then, is the impact of immersion in an emotionally charged environment, and the elicitation of painful, and inspirational stories, and the telling of these stories. Researchers are taught to question their ability, even their right, to use themselves fully as the principal, perhaps only, research instrument.[3] Yet, to know the phenomenon about which they write and to be fully honest about how they came to their interpretation, one can argue that it is dishonest *not* to draw on their own emotional experience and incorporate those emotions into the final telling of their "research tale."

Contributors to this volume were surprised by the emotional impact of their first exposure to the qualitative research process. Certainly, I was surprised by the dreams I experienced as well as my ongoing rumination on the stories of the parents I interviewed.[7,8] This clearly is not unique to this group of people, as it has been reported elsewhere.[10,20] Of importance is that none of us ultimately saw this as a negative experience, even when it had been intense. Our research efforts were enriched by our personal and emotional engagement in the research process.

Qualitative research focuses on the understanding and exploration of sometimes emotionally laden human phenomena. Studies of loss, death, rape, abuse, illness, and endangerment of oneself or a loved one are only a few examples. Rather than objectively reporting observable aspects of the phenomenon, qualitative researchers attempt to enter the subjective world of the researched. In doing this, they report on emotional as well as cognitive elements of some aspect of the lives of those studied. Their understanding of these elements requires empathy, the ability to connect at a feeling and a thinking level with the study participants. Researchers must draw on rational understanding while they also reach within themselves for their subjective views and personal experiences, looking for comparability of experience. Throughout the process, their reactions shape the direction and depth of their understandings of the lives of study participants. As stated before and in sum, the researcher, in effect, becomes the research instrument.

Finding the balance of emotional engagement

To date, attention paid to emotions in qualitative research largely has focused on protecting participants from negative effects. It has not been on doing the same for researchers,[20] or on positive effects. In many cases, researchers have found similar ways to deal with their emotions: they may engage in such reflexive activities as journaling to monitor their personal responses to the research activity.[11,20] Journals that are distinctly different from fieldnotes allow them to keep track of theoretical perspectives, preconceptions and assumptions, as well as emotional reactions.[20] They may seek out social and emotional support.[11] Interviews may be scheduled with time in between so that they can reduce the overwhelming intensity of face-to-face interviews.[21] Rowling[20] suggests developing a sense of "being alongside," avoiding the sense of adopting temporary friendships, but rather being empathic with the ability to "feel" with the other, all the while maintaining a clear internal sense of difference from the other.

In most instances, it is essential that researchers maintain a reflexive stance. Indeed, unless the researcher is self-reflective, a danger exists where the researcher's private affective meanings may cloud understanding of the participants' construction of meaning and eventually complicate the research relationship. At the same time, if researchers distance themselves too much, they run the risk of missing important elements of what they are studying. Thus, they may choose to allow themselves to have certain feelings, but deny or abandon emotions they see as inappropriate to the task at hand.[3]

The relationship with participants

The uncertainty of where appropriate emotional boundaries should lie is an ongoing concern in qualitative research and one that has received surprisingly little published attention, particularly from the lived perspective of the researcher. This is true, even in fields that have a more mature qualitative tradition than psychology.

As indicated earlier, topics explored in qualitative research projects can be highly emotional for both the researched and researcher. Researchers may choose topics that are emotionally close to them, may discover that the focus of their study has a greater emotional impact than they had anticipated, or may find themselves directly affected by whatever it is they are studying. This experience of an emotional as well as an intellectual involvement may be particularly surprising for researchers trained in traditional, positivist approaches that reward emotional distance. The combination of highly charged topics, an in-depth and long-term contact with the phenomenon, and the evolving emotional environment of the researcher's own social world may result in a lack of clarity or "fuzziness" in boundaries. These boundaries must be negotiated and renegotiated, an ongoing part of the research process, as a balance is sought between the dangers and benefits of being too far in or too far out of the lives of the researched.

There may be additional complications regarding boundaries for clinicians conducting qualitative research.[10,20,22,23] For those trained as clinicians accustomed to working with single clients and families on an ongoing basis, the need to remain immersed in the phenomenon over time, but not engaged in a therapeutic relationship with the same clients, may cause them to feel scattered and caught between roles.[22] In addition, ethical guidelines of professional organizations advocate against dual-role relationships, and therapists may feel pulled between their role of therapist and that of a researcher.

Harris and Huntington (Chapter 7) argue for caution in adopting counseling techniques in qualitative research. Stuhlmiller (Chapter 4), on the other hand, describes ways in which qualitative research interviews may actually have a therapeutic effect on the participants who feel empowered by their participation rather than vulnerable, as they might have felt in a client-therapist relationship. Gale[24] echoes this latter view.

Throughout this volume, the authors argue that they have a responsibility to attend to the well-being of others in the process. In my chapter on the emotional impact of qualitative research on staff and students, I write of the project director's role of mentor to the neophyte researcher who is often learning about and experiencing this form of research for the first time. In Chapter 6, Rosenblatt writes of the impact of his research on his family and his efforts to protect his sons from any overwhelming effects. All chapters, to some extent, have referred to the need to watch for the emotional impact on participants in the research. This is done by maintaining an interview structure that allows the participants to work into the process, through the more challenging aspects of the interview, to some sort of closure (see Stuhlmiller for a discussion of this type of structure). In my own work, I label these three segments as the opening and orientation, the body of the interview, and the close and debriefing.

Rowling[20] and others argue for a set of ethical guidelines that specifically address the use of counseling techniques to elicit richly detailed qualitative interviews. It is also true that, as Mienczakowski and his co-authors argue in Chapter 10, we must be aware not only of the emotional impact on the research participants, but also on the audience members and the participants in our various project teams. Perhaps a set of ethical guidelines intended to protect the vulnerable along with an acknowledgment of the reality and, indeed, the appropriateness of emotions in qualitative research is in order.

Final thoughts

This chapter serves as the entry point to this volume on the emotional nature of qualitative research. Each author was given the opportunity to write about how he or she views emotions in qualitative research. Although there was a wide variation in the comfort level in writing about this, they expressed a willingness to tell their stories in a way that legitimized the place of emotion in research, while also maintaining a sense of the appropriateness of research as a well-ordered, intellectual exploration.

Ultimately, we need to recognize and incorporate the emotional nature of qualitative research in our training and education of those new to this research tradition. Doing so will prepare researchers better for their first encounters in the field and move them along the journey that is qualitative research.

References

1. Ellis, C., Emotional sociology. *Studies in Symbolic Interaction*, 12, 123–145, 1991.
2. Reinharz, S., Experiential analysis: A contribution to feminist research, in *Theories of Women's Studies*, Bowles G. and Duelli Klein, R., Eds., Routledge and Kagen Paul, Boston, 1983.
3. Kleinman, S. and Copp, M. A., *Emotions and Fieldwork*, Sage, Newbury Park, NY, 1993.
4. Katz-Rothman, B., Reflection: On hard work, *Qualitative Sociology*, 9, 48–53, 1986.
5. Glaser, B. and Strauss, A., *The Discovery of Grounded Theory*, Aldine, Chicago, 1967.
6. Strauss, A. and Corbin, J., *Basics of Qualitative Research: Grounded Theory Procedures and Techniques*, Sage, Newbury Park, CA, 1990.
7. Gilbert, K. R., Interactive Grief and Coping in the Marital Dyad Following the Fetal or Infant Death of Their Child (Doctoral dissertation, Purdue University, W. Lafayette, Indiana, 1987), *Dissertation Abstracts International*, 49/04A, 962, 1988.
8. Gilbert, K. R. and Smart, L., *Coping with Fetal and Infant Loss: The Couple's Healing Process*, Brunner/Mazel, New York, 1992.
9. Lofland, J. and Lofland, L. H., *Analyzing Social Settings*, McGraw Hill, New York, 1986.
10. Etherington, K., The counsellor as researcher: Boundary issues and critical dilemmas, *British Journal of Guidance and Counselling*, 24, 338–345, 1996.
11. Ely, M., Ed., *Doing Qualitative Research: Circles Within Circles*, Falmer Press, London, 1991.
12. Hochscheild, A. R., The sociology of feeling and emotion: Selected possibilities, in *Another Voice: Feminist Perspectives in Social Life and Social Science*, Millman, M. and Kanter, R. M., Eds., Anchor Books, New York, 1975, 280–307.
13. Oakley, A., Interviewing women: A contradiction in terms, in *Doing Feminist Research*, Roberts, H., Ed., Routledge and Kegan Paul, Boston, 1981, 30–61.
14. Gilbert, K. R. and Schmid, K., Bringing our emotions out of the closet: acknowledging the place of emotion in qualitative research, paper presented at the Theory and Method Preconference Workshop, Annual Meeting of the National Council on Family Relations, Baltimore, Maryland, November 1993.
15. Gilbert, K. R. and Schmid, K., Bringing our emotions out of the closet: Acknowledging the place of emotions in qualitative research, *Qualitative Family Research Network Newsletter*, 8, 4–6, 1994.
16. Jagger, A. M., Love and knowledge: Emotion in feminist epistemology, in *Women and Reason*, Harvey, E. D. and Okruhlik, K., Eds., University of Michigan Press, Ann Arbor, MI, 1992, 115-142.
17. Cacioppo, J. T. and Gardner, W. L., Emotion, *Annual Review of Psychology*, 191, 1999. (online-ISSN: 0066-4308. Available: Northern Lights, www.northern-light.com).
18. Sarbin, T. R., Emotion and act: Roles and rhetoric, in *The Social Construction of Emotions*, Harré, R., Ed., Blackwell, Oxford, 1986, 83–97.
19. Rosenberg, M., Reflexivity and emotions, *Social Psychology Quarterly*, 53, 3–12, 1990.

19a. Goleman, D., *Emotional Intelligence: Why It Can Matter More than IQ for Character, Health and Lifelong Achievement,* Bantam Books, New York, 1995.

20. Rowling, L., Being in, being out, being with: Affect and the role of the qualitative researcher in loss and grief research, *Mortality,* 4, 165–182, 1999.

21. Cook, A. S. and Bosley, G., The experience of participating in bereavement research, *Death Studies,* 19, 157–170, 1995.

22. Hart, N. and Crawford-Wright, A., Research as therapy, therapy as research: Ethical dilemmas in new-paradigm research, *British Journal of Guidance and Counseling,* 27, 205–214, 1999.

23. McLeod, J., Qualitative approaches to research in counselling and psychotherapy: Issues and challenges, *British Journal of Guidance and Counselling,* 24, 309–316, 1996.

24. Gale, J., When research interviews are more therapeutic than therapy interviews, *The Qualitative Report,* 1(4), 31–38, 1993.

chapter two

Feminist research with women awaiting trial: the effects on participants in the qualitative research process

Emma Wincup

Contents

Introduction

The purpose of this chapter is twofold: (1) to explore the impact of engaging in qualitative research on the researcher, and (2) to explore the effects of participating in the research on the individual. The key concern of this chapter is to consider the centrality of emotions in qualitative research, especially in projects informed by feminist theory. This is addressed through a critical review of debates about feminist research and a reflexive account of a three-year

0-8493-2075-5/01/$0.00+$.50
© 2001 by CRC Press LLC

research project on bail hostel* provisions for women awaiting trial in England and Wales. First, I offer an account of a personal journey through the research process and explore the disquiet, discomfort, and self-questioning I experienced as I grappled with the emotional nature of fieldwork.

My first foray into qualitative research was as a doctoral student in the school of social sciences at Cardiff University, Wales. I spent the first months of study familiarizing myself with the methodological literature and with a program of research training. Reading the methodological literature, I reached the conclusion that researchers do not have feelings. It was clear that personal feelings relating to the conduct of fieldwork are usually either hidden or confined to appendices of research monographs or collections of "natural histories" written by experienced researchers.[1-4] These confessional accounts add to textbook prescriptions of how to do research by discussing the mess, confusion, and complexities of actually doing social research. Often, however, the accounts are presented as heroic tales of how methodological problems are confronted and resolved. Moreover, by separating the personal side of fieldwork from the substantive story, they add to the myth that personal feelings do not influence the research to any great extent and do not taint the final product of the monograph, journal article, or report.

I looked to postgraduate seminars for honest discussions of the lived experiences of conducting research. All too often, lecturers discussed methodological issues at an abstract level, with only brief reflections on their experiences in the field. This reflects a pervasive and insidious discourse in social science research: that feelings are equated with weakness and are a barrier to *good* research.[5] Possible reasons for this are now explored.

Traditionally positivist approaches have dominated social research. This form of research emphasizes objectivity in an attempt to establish the discipline of sociology as a science. Consequently, from this perspective, the researcher must adopt a detached stance and ignore the impact of participation in social research. These mainstream (and some feminists would argue "malestream") approaches gloss over the emotions of the researcher and those studied. In contrast, researchers utilizing qualitative techniques have argued for the need for reflexivity to consider the role of values, social processes, and personal characteristics in shaping social research. Qualitative researchers recognize that they are part of the research process and what they observe, hear, and experience is filtered through various lenses including the emotional.

Qualitative researchers, however, receive mixed messages. As they are told to work to establish rapport, they are also told to avoid over-identification and unnecessary emotion. Although qualitative researchers abandon the positivistic goal of objectivity, they still argue for the need to maintain "social distance" to allow analytic work to be accomplished. Thus, many

---

* The term *bail hostel* is used to describe small community-based institutions managed by the probation service for those who have been charged with criminal offenses but have not yet been tried by the courts. Their officially defined role is to provide a supportive and structured environment for those at risk of being imprisoned before trial.

qualitative researchers function like quasi-positivists, allowing themselves to have particular feelings such as closeness with participants but then denying their emotions when they construct their accounts. In other words, researchers engage in emotion work,[6] molding their feelings to meet the expectations of others. Yet, although hidden and unarticulated, emotions continue to influence the research process.

Emotions and feminist research

Since the 1970s, sophisticated debates within feminism have explored the notions of feminist methodology and feminist research practice.[7-10] Of central importance is the need to focus on the complexities and problems of women's situations and the institutions that influence those situations.[8] A variety of approaches, methods, and epistemologies have emerged. One common theme is the rejection of distance and objectivity in the researcher–researched relationship. This has opened up the possibility of focusing on the emotional dimensions of research, which has not been fully exploited.

While some feminists have acknowledged the importance of emotions in debates about ontology and epistemology,[9,10] this has not filtered through to an exploration of how emotions affect the research process and how they can be used as an analytic tool. Thus, despite major shifts in social science research, emotionality is still constructed in opposition to rationality and professionalism, and the importance of emotions is denied. Consequently, talking about the personal impact of conducting research is a task with which many academics (including feminists) are uncomfortable and therefore avoid. This affects the novice researcher by leaving him or her unprepared for the level of emotional engagement that social research requires.

The notion of a research method that is exclusively used by feminists has been challenged by researchers both in and out of the feminist tradition. It is a commonly held view that feminist researchers do not have a blueprint to follow when designing their research projects,[11] or a research method peculiar to feminists. Instead, as Kelly[12] argues, feminists need to use methods that are appropriate to the research questions and that avoid methodological purism. The issue then becomes how to apply a feminist perspective to the research methods employed. However, a number of key themes can be delineated within research projects influenced by feminist theory. Gelsthorpe[13] suggests common themes include choosing topics that are relevant to women and the women's movement; a preference but not an exclusive focus on qualitative research; a reflexive approach, particularly to issues of power and control; and a concern to record the subjective experiences of doing research.

Unlike the position adopted by earlier feminist researchers such as Graham,[14] the choice of a qualitative approach is not based on the premise that only qualitative approaches can be used to generate the kinds of knowledge that feminists wish to develop, and that only qualitative approaches are compatible with the politics of feminism. This viewpoint has been developed from a critique of quantitative approaches which were seen to represent *masculinist* forms of knowledge, emphasizing the objectivity and detachment of the researcher. I do not hold this view. As others have also noted,[12,15,16] research involving quantification make important contributions to our knowledge and understanding of women's lives; for example, through raising awareness of the nature and extent of violence against women.

Feminist researchers do not simply use preexisting research techniques, rather, they adapt them to mesh with their gender-conscious theoretical position. For instance, while feminist researchers have appreciated the compatibility between qualitative approaches to interviewing and feminist concerns, they have also been highly critical about the ways in which traditional textbook guidelines seem to constrain researchers[17,18] because such texts offer a depersonalized and mechanical view of the qualitative research process. In such texts, the interview is seen as forming a hierarchical relationship between the researcher and the researched in which a rapport is only established as a means to gain richer data, and emotional attachment should be avoided at all costs. Instead, feminists advocated a less exploitative relationship based on informality, equality, reciprocity, empathy, rapport, and subjectivity.

There is now recognition that some of these ideals of feminist researchers are difficult to establish in practice. The view that research can be based on a nonhierarchical relationship has been challenged in the light of recent feminist debates. These debates have questioned the essentialist foundations of feminism and drawn attention to the differences that divide women particularly ethnicity, social class, age, and sexual orientation.[19-21] Consequently, the debate has moved on from an uncritical acceptance that feminists can adapt research methods so they are fully congruent with feminist concerns, to a stance which argues for the need to consider the potentials and dilemmas of methods and methodological strategies used in feminist research. In this way, a number of feminist researchers, including myself, have critically reflected on the use of ethnography,[22-24] interviews,[25,26] and documents,[27,28] and their compatibility with feminist principles.

Over the past decade, a great deal has been published on feminist research.[4,29,30] Although others would disagree,[31,32] there has been remarkably little consideration of the role of emotions in the research process. There are a number of possible explanations for this. First, all too often, feminist research has been portrayed as potentially empowering and thus a comfortable and cosy activity. This has often led to the silencing of discussions about the emotional consequences for the women who tell their stories and the researcher who listens. This is remarkable given the sensitive nature of many of the topics selected by feminist researchers, certainly within my own discipline of criminology where feminist researchers have explored sensitive topics such as women's experiences of victimization. The neglect of emotions is also surprising as feminists place a great deal of emphasis on developing a rapport and a close relationship with research *subjects* in order to further the understanding of the problem being investigated.

Second, women who adopt a feminist approach face a major dilemma in that they may get caught up in a vicious circle. If they explore emotional issues, they fit a stigmatized stereotype because women are often assumed to be emotional beings. If they marginalize such issues, they perpetuate the myth that emotions do not influence the research process. Although there has been a growing recognition that women need to locate themselves within their research and writing,[29,32a] these ways of working are often criticized because they represent challenges to the traditional academic approach. Inclusion of the self may lead to the criticism that one's work is not *academic* enough and, even worse, mere self-indulgence.[33] Given that academic work is subjected to criticism through the peer reviewing process, critical comments may lead to greater feelings of distress because they may be interpreted as criticism of the self, as well as the work produced.[33]

Feminist concerns about the emotional effects on participants in the research process have only recently begun to be voiced. Maynard and Purvis[4] focused on themes and controversies in feminist research, including the impact of involvement in research which is a welcomed addition to methodological debates. Phoenix[34] explored the impact on the researcher when the research involves exposure to racism accounts of gross humiliation and torture,[28] and revelations of violence and abuse.[35] Another dimension explored was the impact on those studied. In this context, Phoenix[34] reported on two studies of young women that asked for their views of and responses to the research. Phoenix noted that although many women said they enjoyed being interviewed, they raised concerns relating to the intrusiveness of the questions and the length of time the interview had taken. As did other contributors,[22,35] Phoenix signaled the necessity for feminists to consider further the likely impact on those who participate in the research process.

There are many questions still to be fully explored:

- What are the effects of asking women to talk about painful aspects of their personal lives?
- How do participants experience the research process?
- Do participants see the research process as empowering, distressing, intrusive, or in some other way?
- What are the emotional costs to the researcher of conducting qualitative research on women's lives?

These questions are considered in the following discussion which offers a reflexive account of conducting feminist research with women awaiting trial.

Researching women's lives

The initial choice of topic was partially an emotional decision. An important advice was given to me as a postgraduate student studying for a master's degree in criminology: choose a Ph.D. thesis topic that makes you angry. The rationale for this was that anger would help to sustain my interest through three years of individual and relatively isolated study. At that time, I began to read a number of autobiographical and highly critical accounts of women's imprisonment.[36-38] All made me angry (and they still do) but one in particular held my attention: Audrey Peckham's[38] vivid account of awaiting trial. Her descriptions of the humiliations and degradations of waiting anxiously for trial in a remand center,* where she was denied access to appropriate psychiatric care, were distressing. Her suggestion that her treatment as an unconvicted prisoner was worse than her treatment as a convicted prisoner was particularly disturbing. Peckham's account concluded with a suggestion that, where possible, those awaiting trial should be diverted from the prison system to supportive institutions such as bail hostels.

Unsure about the role of bail hostels in the criminal justice system but having read similar suggestions by penal pressure groups and academic criminologists, I selected the study of bail hostels as alternatives to women's imprisonment as my research topic. More specifically, the research aimed to answer three questions: (1) what are the particular problems experienced by women awaiting trial, (2) how do bail hostels attempt to offer support, and (3) what are the factors that help or hinder the provision of support? This research involved qualitative fieldwork in two stages: first, in a remand center and second, in three bail hostels

The research project was influenced by feminist perspectives in criminology. Feminist critiques of academic disciplines have been influential in reorienting research agendas to topics that are relevant to women's lives. Within criminology, a "patchwork"[39] of knowledge has been developed by feminist researchers on women's experiences as victims, offenders and, to a lesser extent, criminal justice professionals. Such researchers have exposed the neglect and misrepresentation of women in traditional criminological research.[39a,39b] The choice of a feminist theoretical framework influenced the framing of research questions, what data are collected in order to answer those questions, and how the data is categorized and subsequently analyzed. Feminist theory is drawn upon rather than imposed.[29] In other words, the use of feminist theory had important implications for the design and conduct of the research, but did not rigidly constrain the research or prohibit the use of theoretical concepts from critical and mainstream criminology and sociology.

Qualitative approaches appeared the most congruent with the aims of the research because such approaches are particularly suited to exploratory research that attaches an importance to context, setting, and the participant's frame of reference.[39c] Qualitative data can be collected using a range of methodological techniques, such as interviews and observation, and presented to a future audience through meaningful descriptions of how social life is accomplished. Ethnography — defined by Fetterman[40] as "the art and science of describing a group or culture" — was selected as an appropriate methodological strategy because it allows "an investigator to establish a many-sided and relatively long-term relationship"[41] with groups of individuals

---

* A remand center is a custodial institution for those charged with criminal offenses but not yet tried. Often remand centers are housed in a prison.

in a natural setting. Specifically, the adoption of an ethnographic approach allowed empathic understanding of criminal justice institutions for women and the lives of the women who live and work in them. Ethnography involves watching what happens, listening to what is said, and asking questions; in other words, it resembles the ways in which people make sense of their everyday lives. This has been regarded as a fundamental strength of ethnography for many social scientists.[42]

Although often used as a term that is synonymous with participant observation, ethnography is a methodological strategy used to research people's lives for an extended period of time in their own surroundings through participant observation techniques in conjunction with other methods. Contemporary ethnography tends to be multi-method research.[29,43,44] In keeping with this tendency, my own research involved a triangulation of methods including participant observation, interviews, and documentary analysis.

Over a period of 18 months, I spent long periods of time with women who were facing criminal charges. I negotiated access to a remand center by writing to the prison governor and spent six months visiting on a weekly basis. Initially, I worked as a volunteer in the education department and later spent time in other areas of the prison, such as the accommodation areas. The importance of this initial fieldwork was twofold: (1) it provided me with an opportunity to familiarize myself with women's prisons before commencing research on alternatives to prisons, and (2) it served as a pilot stage of ethnographic techniques. Later in this chapter, I explore how this sent me on a steep learning curve on emotions and social research by introducing me to the challenging and stressful experience of researching the lives of women awaiting trial.

The next stage was to undertake the main body of the fieldwork in bail hostels for women awaiting trial. Following a lengthy series of negotiations, letters, and phone calls to chief probation officers and hostel managers, the three hostels I approached agreed to participate in the research. My sample comprised a women-only hostel, a women-only hostel with provision for children, and a mixed hostel that accommodated mainly men but had a small annex with four beds reserved for women. In these hostels, I interviewed 15 female hostel workers and 15 residents, although in this chapter I reflect largely on the interviews with the residents. The interviews with the hostel workers played a crucial role in developing an understanding of the treatment of women and allowed examination of the constraints in which their work with women awaiting trial is carried out.[44a]

The interviews were semi-structured in format. Straddling the divide between "standardized" and "reflexive" interviewing,[42] semi-structured interviews allow women to introduce issues they would like to discuss and therefore help interviewers gain insight into the most important aspect of women's lives. The use of an interview schedule provides the interviewer with a clear agenda, allowing particular questions to be asked to ensure comparability and to facilitate data analysis by identifying some initial themes. The interviews were the main source of data for the research project but were supplemented by the use of two other research methods: participant observation and documentary research.

I spent long periods of time participating in and observing the everyday activities of bail hostels such as: attending staff and resident meetings, watching television with residents, sharing meals and cups of coffee, shadowing staff, and chatting with residents and visiting professionals as they came in and out of the main office — the hub of hostel activity. I wrote my observations in fieldnotes which included descriptions of social processes and their contexts, including the emotional context — for example, my own feelings and emotional responses to participants — through which what I observed and experienced as reality was filtered. In addition, a variety of documents and records were available to me as a researcher: log books in which the staff recorded significant events that had happened in the day, publicity brochures, rules, regulations and policies, as well as official documents produced by individual probation services and the government, such as plans and strategies.

A personal and emotional journey

I struggled to cope with the emotional engagement necessary to conduct ethnographic research with women experiencing crisis. In this instance, they were awaiting trial, often for serious criminal charges. They were also experiencing multiple complex problems, including the effects of abuse (emotional, sexual, and physical), poverty, homelessness, unemployment, sole responsibility for dependants, and substance misuse.[45-47] Prior to the fieldwork, I had not sufficiently reflected upon the effects of the research on the participants in the research process, including myself. Coupled with the lack of self-confidence shared by many doctoral students, I began thinking that my approach to research was inadequate in some way. Had I begun to get too involved? Had I "gone native"? Gradually I discovered that other social science researchers experience similar feelings, yet often those feelings remain hidden.

My fieldwork took me away from the university for long periods of time and away from the support networks of supervisors and fellow students. I felt isolated as I battled with the emotional consequences of conducting research and my own anxieties about the effects of the research on the participants. A landmark for me was meeting congenial colleagues at a summer school for doctoral students a few months into the fieldwork. Here we articulated our feelings about conducting research — many of us for the first time. I began to recognize my own feelings as more than research inadequacies and began to take them seriously. One year later, the inclusion of papers focusing on emotions and qualitative research at a conference largely dominated by papers on quantitative methods demonstrated the importance of this neglected issue.

Safe in the knowledge that others have experienced similar emotions and that I now have my Ph.D. and have embarked on a relatively successful research career, I can write about feelings evoked during fieldwork. I can now view feelings as a source of strength, not weakness, which allows me to develop a truly reflexive and honest account of the research process. Much of this chapter is a confessional tale. Before telling my story, I offer an overview of the research design and discuss the theoretical and methodological debates that influenced the research.

Other researchers have commented on their early days in the field.[48,49] From my own experience, they were anxious times. I had no experience of working in criminal justice agencies and no contact with female offenders until I went into the remand center. I felt uneasy in an unfamiliar setting. I felt apprehensive being surrounded by individuals accused, in some cases, of very serious offenses, but this stemmed from my inability to predict their reactions toward me rather than from a fear of victimization. I wondered whether they would regard me as an outsider who might have difficulty understanding their situations and experiences, and whether they would perceive what I was trying to achieve as worthwhile. I felt apprehension about their responses to me and their cooperation in the research process. Anxious thoughts entered my head. I wondered, would they find my presence too intrusive? Would they want to talk to a stranger about their experiences? Given the women's histories of institutional confinement and other formal interventions in their lives, the expectation that they would be wary of my inquiries seemed reasonable. This was especially the case given the personal and painful nature of the subject matter of our discussions.

Despite my concerns, 15 women awaiting trial agreed to participate in the research by being interviewed, and I spoke to many others in a more informal way. It would be wrong for me to give the impression that I easily found women willing to be interviewed. Some of the women refused when I asked them to participate. One possible reason for this might be that some women felt that too many other things were going on in their lives at the time. Obviously some people are more open about their lives than others. This latter explanation was frequently mentioned in interviews with members of staff when I asked them to reflect on their own interactions with the women.

Many early discussions around feminist methodology[17,18] were concerned with interview dynamics. Critical feminist methodological comments highlighted that interviews should involve an exchange of information. Even in the early stages of the project, I sensed that those I spoke with wanted more than to answer my questions and bombarded me with their own questions. One comment from a woman in a remand center brought this to light: "I'd be interested to hear your views unless you've only come here to collect information." I assured her that I had not, and we began a conversation about many things including decriminalization of soft drugs (she had been charged with dealing cannabis which she maintained she supplied only for medicinal purposes). Whether her comment was a plea for emotional involvement as feminists suggest or a request for information from a credible source, I do not know. Crucially, it made me reflect on my developing personal involvement in the project.

In the bail hostels I visited, staff members frequently asked me questions about what other hostels provided. Hostels are managed by local probation services working within national guidelines. Consequently, they had few opportunities to meet workers from hostels in other areas, so they saw me as someone who could share information about good and bad practice. The notion of reciprocity in interviews with the women residents was more problematic. I was conscious of my limitations. I was aware that some of the women would simply want someone to talk to and that I could provide this. However, I was not a trained counselor, therapist, or medical practitioner.

In particular, I was wary about offering reassurances on future sentencing outcomes. In one case, a woman told me that she was terrified of going back to prison (she had spent a short time on remand in Holloway, London) and asked me how likely it was she would get a probation order. I hoped that as a young single parent charged with her first offense she would not be imprisoned. I could not be certain of this however, especially as she was charged with a serious violent offense. The research took place at a time when the political context emphasized that "prison works", and the prison population was rapidly rising. Hence, I carried on listening to her fears, respected them as very real fears, and tried to answer factual questions such as: what is a probation order?* At a professional level, I knew this was the right thing to do, but at a personal level I felt a desire to offer reassurances as one might for a friend. I found myself pulled in two directions about the best way to proceed.

My own biography was crucial to the research. The women I interviewed who were awaiting trial were a diverse group in terms of age, social class, ethnicity, nationality, and sexual orientation and they had life experiences that were very different from my own. Throughout, I was acutely aware of the differences between my own relatively privileged lifestyle and that of many of the women I met. The women I spoke to were also very aware of these differences. The very fact that I had chosen to spend time in a criminal justice institution rather than be required to live in one created a barrier between us. While I could walk away at the end of the day, their freedom was curtailed by the courts.

So, establishing rapport and empathy was not inevitable because I was a woman researching women's lives, and a sense of "shared sisterhood" was not predestined. While I could identify with at least some of the women and some of their experiences, I could never put myself in their shoes. Regardless of how much they informed me of their lives, I could not escape my own

---

* A probation order is a sentence issued by the criminal courts. It is an alternative to custody. It requires the offender to be supervised by a probation officer who meets with him or her regularly. Other conditions may be attached, for example, attendance at group work programs for drug and alcohol abuse.

personal history. Differences between us were usually more apparent than similarities (see Reference 50 for a similar reflexive account of researching women prisoners). My emotional responses to this realization were mixed. I felt isolated as a result of my difference from them, coupled with feelings of relief that I had not shared their life experiences. Other researchers have noted similar feelings (see Chapter 7).

Research relationships with women awaiting trial had to be worked at to overcome the apprehension of both parties. My fears, discussed above, were coupled with a feeling held by some of the women awaiting trial that they had nothing of interest that they could tell me as a researcher. Given these preconceptions, often both the interviewee and I were surprised as the interview unfolded, and rich and detailed accounts were offered. From my point of view these offered fascinating data. My understanding of the women's views of the research experience is explored later in this chapter.

The impact of the research: a personal view

The fieldwork proved to be a distressing experience for me. Through interviews I heard stories of domestic violence, poverty, drug and alcohol dependency, and abusive experiences and learned something of the difficulties of waiting for trial. Through observation, I saw firsthand the impact of substance misuse, women with injuries caused by violence, and the depression, anxiety and anger associated with waiting for trial. A particularly disturbing experience for me occurred early in my fieldwork at the remand center. A woman approached and held in front of me her self-mutilated wrists. While I was aware that self-harm is frequent in women's prisons, especially among the remand population,[51] confronting with the stark reality of it was another matter. I experienced a number of emotions, largely a mixture of shock and sadness. Certainly, I had not anticipated the willingness of some women to share sensitive, personal, and often painful aspects of their lives.

Some of the women I spoke with had reached a point in their lives where they had a need and a desire to talk about their experiences, and the interviews gave them an opportunity to tell their stories. The semistructured interview included questions on sensitive topics such as experiences of abuse, but often clearly painful experiences tumbled out with no prompting. Like Parr,[25] I experienced conflicting emotions: sadness and distress at the women's experiences, but privilege that they trusted me with their stories. I was acutely aware that some of the women were telling me things that they may never have revealed before.

Reflecting on my own research experiences, I feel it is important to problematize the view that research relationships are always built on empathy where the participants are able to identify with each other. This is particularly important in my own discipline. Few criminologists have spoken honestly about their feelings in their research on offenders. A significant exception is Adams'[52] exploration of the fluid emotions of love and hate when researching the pretrial process accounts of the suspects:

> On some days, these interviews were exhilarating and
> informative, I felt I really understood the suspects'
> "plight": I loved 'em and my research, they were boys
> not men, they had gone astray, they needed under-
> standing and protection from further harm. On other
> days, I felt angry and annoyed and could not under-
> stand why some suspects whinged incessantly about
> poor treatment from the police, especially in the face
> of what they had done to others: I hated 'em and my
> research, they deserved what they got.

This unwillingness to be honest about negative feelings toward participants
is not peculiar to criminology researchers. There are few accounts in which
authors have been open about feelings such as anger towards participants.[53]

Although I established rapport and empathy with some women and
formed close research relationships with others, I also had negative feelings
toward some during the fieldwork. Most women who appeared in court
were charged with minor property offenses;[54] however, I met women
charged with crimes of violence, child abuse, and murder. I also saw women
bullying and victimizing other women. I did not judge them, or at least I
tried not to. Instead, I attempted to understand why they had done what
they had done, even if I could not condone their actions.

A particularly difficult situation arose in the remand center (and a similar
situation arose in the first hostel I studied). Other prisoners discovered that
one of the women had been charged with sexual offenses against children,
and this resulted in the woman being ostracized and treated with contempt.
This was the woman who had previously showed me her self-mutilated
wrists. While I could not excuse the actions of the women, I could understand
why they acted in the way they did. Many were mothers whose children
were being cared for by family members or by foster parents, or who were
in children's homes while their mothers awaited trial. Many had experienced
abuse themselves. As a result, this was a situation they found deeply distressing and a source of great anxiety. Consequently, they found it hard to
cope when confronted by this woman, although she had yet to be convicted
of these charges. I did not want to join in the name calling but, at the same
time, I did not want to alienate the other women prisoners by challenging
their view. Instead, I steered the conversation on to other topics.

Fieldwork in stressful and emotionally charged situations affected my
life greatly. I was not prepared for the emotional effect it would have on me.
Practitioners who confront these issues on a day-to-day basis receive, at least
in theory, support in the form of supervision and staff meetings. I explored
the dimensions of formal and informal support in interviews with hostel
workers. However, as a researcher I did not have this support and instead
used the safety valves suggested by Glesne and Peshkin:[55] taking a break
(strong words in a fieldwork diary) and becoming involved in other research
activities. Largely these could be described as "sticky plaster" or "Band-Aid®"

techniques, means of hiding the issues presented. They are no substitute for discussing feelings associated with fieldwork with other researchers. Peer discussion can provide reassurance and helps to overcome feelings of isolation by recognizing that your own emotional experiences are not unique.

Even after the fieldwork had ended my emotional involvement continued. Leaving the field physically was relatively easy. I had made it clear from the outset that my presence would be for a short time. Getting away from the field emotionally proved more challenging. I found it too hard to distance myself from the settings, and the data analysis process served as a constant reminder of who I had met, the problems they had experienced, and the emotions they displayed when telling their stories. I relived the interviews when transcribing the tapes. The interviews included questions about the women's future hopes and aspirations, and led to great curiosity about their lives after leaving the hostel. Now, three years after receiving my Ph.D. and more than four years after completing the fieldwork, I ask myself why I ever wanted to distance myself from the settings.

On reflection, my attempts at emotional disengagement were in part a way of coping with the impact of fieldwork as I continued with the stages of "data analysis" and "writing up." These terms frequently used by researchers are not simply stages in the research process; they relate to people's lives. My data were the stories of women's experiences. My writing was more than creating a Ph.D. thesis. I was creating an account of women's experiences through my interpretation of their voices. Gradually, I began to realize that my own emotions could be used constructively to enhance the research without compromising the ability to step back and offer analysis and interpretation. Reliving my fieldwork experiences brought my data alive. My emotional awareness encouraged me to listen more closely to the accounts of the women I interviewed, and to think carefully about the ways I could do justice to the women's stories as I created my ethnographic account.

More harm than good? Unanswered questions about the impact of the research

Participating in the research was also an emotionally charged experience for the women who agreed to be interviewed. The women awaiting trial showed a diversity of emotions when telling their stories. Some became upset and began to cry, others talked in angry and bitter ways about their experiences, while some appeared not to show any emotions at all. The interviews may have been empowering in the sense that they were an opportunity for the women to articulate their experiences in the hope that it might lead to change; or they may have been cathartic, providing an outlet for individuals to off-load. More simply, they may have provided women with an opportunity to talk about their lives to another interested individual.

I tried to convince myself of the benefits of the research discussed above; however, these had to be reconciled with a very real fear that in some ways

I had caused harm, dredging up painful memories for women already experiencing a difficult period in their lives. Was I contributing to the difficulties already experienced by vulnerable and marginalized women? It was certainly not my intention, but I could not rule this out as an unintended consequence of the research. I began to feel that my presence among these women in crisis was intrusive and potentially exploitative. This feeling has been described by other researchers. As Glesne and Peshkin[55] wrote:

> Questions of exploitation ... tend to arise as you become immersed in research and begin to rejoice in the richness of what you are learning. You are thankful, but instead of simply appreciating the gift, you may feel guilty for how much you are receiving and how little you are giving in return.

Asking myself the question of who benefits, I felt that ultimately I was the one to benefit from the research, often resulting in feelings of guilt. I had attended a conference on women and crime where a member of a penal pressure group had spoken angrily about academics building careers on the back of women in prison. I began to wonder whether I was doing the same thing. I was working toward a Ph.D. that would further my career by researching the lives of women whose circumstances I could perhaps never really comprehend, under the pretext of improving the lot of women who appear before the courts.[56]

Being sensitive to my own emotions encouraged me to reflect on the ethical dimensions of researching women in crisis and to consider the extent to which the effects of conducting research and participating in research is adequately discussed in feminist methodological debates. From the design stages on, I had chosen methods compatible with feminist principles. It was my intention to provide a context where women could voice their experiences and to some extent set their own agenda in the interview. However, one outcome of allowing the women to choose the depth of their disclosures (through open-ended questions and possibilities to decline to answer questions) was that some may have felt greater emotionality than they might have in a more structured interview situation. There was a fine line between using methods that enabled the women to speak in as much detail about their experiences as they wished and unintentionally steering the women to experience painful and potentially damaging feelings.

In order to explore this theme further we, as researchers, need to understand the nature of emotions, both positive and negative. A useful way may be to look toward the therapeutic literature for discussion of these issues (see Chapter 7). As I conducted the fieldwork, I automatically assumed that when women began to get upset or cry, it was a negative situation which I should have avoided. This led me to steer the interviewee onto other topics and sometimes cut the interview short. I downplayed the fact that some of the women, including those who had gotten upset, had said they had welcomed the opportunity to talk and enjoyed the interview. Thus, when the women expressed emotions such as sadness, I felt that I was to blame. I was especially anxious because I had been informed by the manager at the first hostel I visited that it was my role to take responsibility for the emotional effects of asking women to discuss their lives.

It is important to acknowledge that the women did have choices about whether to participate in the research, and about how much detail to provide. At the outset, women were given the opportunity not to answer questions that made them uncomfortable. Nonetheless, although I was not the principal cause of a woman's emotions and these emotions expressed were not necessarily negative or unwelcomed, I felt some obligations to the women who had shared their stories with me. Like Eaton,[57] I deliberately concluded the interviews with future hopes so that they ended on a more positive note. Also, interviews often ended with informal conversations, and I made sure I did not walk away leaving the women in a distressed state.

Conclusion

Qualitative research with women awaiting trial presented an emotional challenge. Despite the concerns I have raised in this chapter, I feel that it was an appropriate approach to an understanding of the lives of women awaiting trial. Although I have raised ethical questions, I hope that by telling the women's stories with passion and conviction, I can improve the lot of women awaiting trial. I do not delude myself that research greatly contributes to policy and practice; instead, I hope that the growing body of work on the lives of women who appear before the courts may have a gradual effect on criminal justice policy.

Although some would criticize the subjectivity that is inherent to my work, I feel that sensitivity to my own feelings and those of the participants deepens understanding and enhances the creation of meaning. Being honest about one's own feelings makes explicit how one's accounts are context bound and strengthens one's integrity as a researcher.[58] Keeping them hidden perpetuates the myth that personal feelings do not influence the research process. As Jaggar[59] notes, "lacking awareness of [our] emotional responses frequently results in [our] being more influenced by emotion rather than less." Being sensitive to the impact of research on those we interview or observe encourages us to view them as participants in the research process, and not subjects or respondents or sources of data. Without them, there is no research process. This necessitates that we reflect upon the ethical dimensions of our work. Honest reflections such as these can only enhance qualitative approaches to social research.

My experiences of qualitative research as a doctoral student influence me now in my role of supervisor and convener of postgraduate research methods courses. I encourage students to consider the potential emotional effects when designing their studies, and frequently talk about my own experiences of conducting research (possibly too much). Many of the criminology students I work with will embark on their first experience of qualitative research on challenging and difficult topics. If we take emotions seriously, we need to explore practical strategies to work with our emotional responses. Strategies for helping students cope with negative emotional effects, such as anger or distress, need to be incorporated into teaching, training, and supervision (see Chapter 8) and be combined with an appreciation of how emotions are also a positive aspect of qualitative research.

Acknowledgments

Thanks to Alexander Massey, Chris Powell, and Kathleen Gilbert for commenting on earlier versions of this chapter.

References

1. Bell, C. and Newby, H., Eds., *Doing Sociological Research*, Allen and Unwin, London, 1977.
2. Bell, C. and Roberts, H., Eds., *Social Researching: Politics, Problems and Practice*, Routledge, London, 1984.
3. Hobbs, D. and May, T., Eds., *Interpreting the Field: Accounts of Ethnography*, Oxford University Press, Oxford, 1993.
4. Maynard, M. and Purvis, J., Eds., *Researching Women's Lives from a Feminist Perspective*, Taylor and Francis, London, 1994.
5. May, T., Feelings matter: inverting the hidden equation, in *Interpreting the Field: Accounts of Ethnography*, Hobbs, D. and May, T., Eds., Oxford University Press, Oxford, 1993, 69–97.
6. Hochschild, A., *The Managed Heart: The Commercialization of Human Feeling*, University of California Press, Berkeley, CA, 1983.
7. Harding, S., *Feminism and Methodology*, Open University Press, Buckingham, U.K., 1987.
8. Olesen, V., Feminisms and models of qualitative research, in *The Handbook of Qualitative Research*, Denzin, N. and Lincoln, Y., Eds., Sage, Newbury Park, CA, 1994, 158–174.
9. Stanley, L. and Wise, S., *Breaking out: Feminist Consciousness and Feminist Research*, Routledge and Kegan Paul, London, 1983.
10. Stanley, L. and Wise, S., *Breaking out Again: Feminist Ontology and Epistemology*, Routledge, London, 1993.
11. Gelsthorpe, L. and Morris, A., The transformative experience in feminist research: from practice to theory, in *Feminist Perspectives in Criminology*, Gelsthorpe, L. and Morris, A., Eds., Open University Press, Buckingham, U.K., 1990, 85–88.
12. Kelly, L., Journeying in reverse: Possibilities and problems in feminist research on sexual violence, in *Feminist Perspectives in Criminology*, Gelsthorpe, L. and Morris, A., Eds., Open University Press, Buckingham, U.K., 1990, 107–114.
13. Gelsthorpe, L., Feminist methodologies in criminology: a new approach or old wine in new bottles? in *Feminist Perspectives in Criminology*, Gelsthorpe, L. and Morris, A., Eds., Open University Press, Buckingham, U.K., 1990, 89–106.
14. Graham, H., Surveying through stories, in *Social Researching: Politics, Problems and Practice*, Bell C. and Roberts, H. Eds., Routledge, London, 1984, 104–120.

15. Kelly, L., Regan, L., and Burton, S., Defending the indefensible? Quantitative methods and feminist research, in *Working Out: New Directions in Women's Studies*, Hinds, H., Phoenix, A., and Stacey, J., Eds., Falmer Press, London, 1992, 149–160.

16. Pugh, A., My statistics and feminism — a true story, in *Feminist Praxis: Research, Theory and Epistemology in Feminist Sociology*, Stanley, L., Ed., Routledge, London, 1990, 103–112.

17. Finch, J., It's great to have someone to talk to, in *Social Researching: Politics, Problems and Practice*, Bell, C. and Roberts, H., Eds., Routledge, London, 1984, 70–87.

18. Oakley, A., Interviewing women: A contradiction in terms, in *Doing Feminist Research*, Roberts, H., Ed., Routledge, London, 1981, 30–61.

19. Anthias, F. and Yuval-Davis, A., Contextualizing feminism: Gender, ethnic and class divisions, *Feminist Review*, 5, 62–73, 1983.

20. Hill-Collins, P., *Black Feminist Thought: Knowledge, Consciousness and the Politics of Empowerment*, Unwin Hyman, London, 1990.

21. Spelman, E., *Inessential Woman: Problems of Exclusion in Feminist Thought*, The Women's Press, London, 1990.

22. Skeggs, B., Situating the production of feminist ethnography, in *Researching Women's Lives from a Feminist Perspective*, Maynard, M. and Purvis, J., Eds., Taylor and Francis, London, 1994, 72–92.

23. Stacey, J., Can there be a feminist ethnography?, *Women's Studies International Quarterly*, 11, 21–27, 1988.

24. Wincup, E., Researching women awaiting trial: Dilemmas of feminist ethnography, in *Qualitative Research in Criminology*, Brookman, F., Noaks, L., and Wincup, E., Eds., Ashgate, Aldershot, 1999.

25. Parr, J., Theoretical voices and women's own voices: The stories of mature women students, in *Feminist Dilemmas in Qualitative Research: Public Knowledge and Private Lives*, Ribbens, J. and Edwards, R., Eds., Sage, London, 1998, 87–102.

26. Puwar, N., Reflections on interviewing women MPs, *Sociological Research Online*, 2(1), Available: <http://www.socresonline.org.uk/socresonline/2/1/4.html>.

27. Bell, L., Public and private meanings in diaries: Researching family and childcare, in *Feminist Dilemmas in Qualitative Research: Public Knowledge and Private Lives*, Ribbens J. and Edwards, R., Eds., Sage, London, 1998, 72–84.

28. Purvis, J., Doing Feminist Women's History: Researching the lives of women in the suffragette movement in Edwardian England, in *Researching Women's Lives from a Feminist Perspective*, Maynard, M. and Purvis, J., Eds., Taylor and Francis, London, 1994, 166–189.

29. Reinharz, S., *Feminist Methods in Social Science Research*, Oxford University Press, New York, 1992.

30. Ribben, J. and Edwards, R., *Feminist Dilemmas in Qualitative Research: Public knowledge and private lives*, Sage, London, 1998.

31. Carter, K. and Delamont, S., Introduction, *Qualitative Research: The Emotional Dimension*, Carter, K. and Delamont, S., Eds., Avebury, Aldershot, 1995.

32. Powell, C., Whose voice? Whose feelings? Emotions; the theory and practice of feminist methodology, in *Qualitative Research: The Emotional Dimension*, Carter, K. and Delamont, S., Eds., Avebury, Aldershot, 1995, 49–71.

32a. Stanley, I., *The Auto/Biographical I: The Theory and Practice of Feminist Auto/Biography*, Manchester: Manchester University Press, 1995.

33. Letherby, G., The personal and the collaborative, in British Sociological Association Equality of the Sexes Committee, Eds., *Writing and Publishing*, British Sociological Association, Durham, 1999, 28, 29.

34. Phoenix, A., Practising feminist research: The intersection of gender and 'race' in the research process, in *Researching Women's Lives from a Feminist Perspective*, Maynard, M. and Purvis, J., Eds., Taylor and Francis, London, 1994, 49–71.

35. Holland, J. and Ramazanoglu, C., Coming to conclusions: Power and interpretation in researching young women's sexuality, in *Researching Women's Lives from a Feminist Perspective*, Maynard, M. and Purvis, J., Eds., Taylor and Francis, London, 1994, 125–148.

36. Carlen, P., Ed., *Criminal Women: Autobiographical Accounts*, Polity Press, Cambridge, 1985.

37. Padel, U. and Stevenson, J., *Insiders: Women's Experiences of Imprisonment*, Virago, London, 1988.

38. Peckham, A., *A Woman in Custody*, Fontana, London, 1985.

39. Heidensohn, F., Feminist criminologies: Directions for the future, unpublished paper presented to the Institute of Criminology, Cambridge, England, 1994.

39a. Heidensohn, F., The deviance of women: a critique and an enquiry, *British Journal of Sociology*, 19:160175, 1968.

39b. Smart, C., *Women, Crime and Criminology: A Feminist Critique*, London: Routledge, 1976.

39c. Marshall, C. and Rossman, G., *Designing Qualitative Research*, Sage, Newbury Park, CA, 1989.

40. Fetterman, M., *Ethnography: Step by Step*, Sage, Newbury Park, CA, 1989.

41. Lofland, J. and Lofland, L., *Analysing Social Settings: A guide to Qualitative Observation and Analysis*, Wadsworth, Belmont, CA, 1984.

42. Hammersley, M. and Atkinson, H., *Ethnography: Principles in Practice*, Routledge, London, 1995.

43. Pearson, G., Talking a good fight: Authenticity and distance in the ethnographer's craft, in *Interpreting the Field: Accounts of Ethnography*, Hobbs, D. and May, T., Eds., Oxford University Press, Oxford, 1993, vii–xvii.

44. Schatzman, L. and Strauss, A., *Field Research: Strategies for a Natural Sociology*, Prentice-Hall, Englewood Cliffs, NJ, 1973.

44a. Wincup, E., *Residential Work with Offenders: Reflexive Accounts of Practice*, Aldershot: Ashgate, in press.

45. Wincup, E., *Waiting for trial: Living and working in a bail hostel*, unpublished Ph.D. thesis, University of Wales, Cardiff, 1997.

46. Wincup, E., Power, control and the gendered body, in *The Body in Qualitative Research*, Richardson, J. and Shaw, A., Eds., Ashgate, Aldershot, 1998, 107–125.

47. Wincup, E., Women awaiting trial: common problems and coping strategies, *The British Criminology Conferences: Selected Proceedings*, 2, 1999. [On-line]. Available: <http://www.lboro.ac.uk/departments/ss/bsc/bsccp/vol02/11wincup.htm>.

48. Van Mannen, J., Playing back the tape: Early days in the field, in *Experiencing Fieldwork*, Shaffir, W. and Stebbins, R., Eds., Sage, Newbury Park, CA, 1991, 31–42.

49. Shaffir, W., Managing a convincing self-presentation: Some personal reflections on entering the field, in *Experiencing Fieldwork*, Shaffir, W. and Stebbins, R., Eds., Sage, Newbury Park, CA, 1991, 72–86.

50. Comack, E., Producing feminist knowledge: lessons from women in trouble. *Theoretical Criminology*, 3, 287–306, 1999.
51. Coid, J., Wilkins, B., Coid, B., and Everitt, B., Self-mutilation in female remanded prisoners II: a cluster analytic approach towards the identification of a behavioural syndrome, *Criminal Behaviour and Mental Health*, 2, 1–14, 1992.
52. Adams, C., Suspect data: Arresting research, in *Doing Research on Crime and Justice: Reflections on the Research Process*, King, R. and Wincup, E., Eds., Oxford University Press, Oxford, 2000, 385–394.
53. Kleinman, S. and Copp, M., *Emotions and Fieldwork*, Sage, Newbury Park, CA, 1993.
54. Heidensohn, F., Gender and crime, in *The Oxford Handbook of Criminology*, Maguire, M., Morgan, R., and Reiner, R., Eds., Oxford University Press, Oxford, 1997, 761–798.
55. Glesne, C. and Peshkin, A., *Becoming Qualitative Researchers*, Longman, White Plains, NY, 1992.
56. Williamson, H., Systematic or sentimental? The place of feelings in social research, in *Qualitative Research: The Emotional Dimension*, Carter, K. and Delamont, S., Eds., Avebury, Aldershot, 1995, 28–41.
57. Eaton, M., *Women after Prison*, Open University Press, Buckingham, 1993.
58. Sword, W., Accounting for the presence of self: Reflections on doing qualitative research, *Qualitative Health Research*, 9, 270–278, 1999.
59. Jaggar, A., Love and knowledge: Emotion in feminist epistemology, in *Gender/Body/Knowledge: Feminist Reconstructions of Being and Knowing*, Jaggar, A. and Bordo, S., Eds., Cornell University Press, New Brunswick, NJ, 1989.

chapter three

The management of self: practical and emotional implications of ethnographic work in a public hospital setting

Hilary Davis

Contents

0-8493-2075-5/01/$0.00+$.50
© 2001 by CRC Press LLC

Introduction

Researchers working in hospital settings are likely to encounter illness and sudden death. This may be upsetting or threatening, particularly for the novice researcher. This chapter recalls some of my experiences while doing ethnographic fieldwork in an English public hospital setting. In particular, I explore some of the practical problems and emotional experiences I encountered when working in a variety of acute and general care areas within the hospital. I argue that in hospital settings we need to recognize the constant interplay between the personal, the emotional, and the intellectual work of the researcher, as we strive to achieve a credible story of hospital life. In particular, I draw upon some of the ideas of role theory to highlight the ambiguity of the researcher who plays a (purported) nonparticipant role in this particular semipublic milieu. Ultimately, I argue that we need to recognize the role of emotional labor for both the researcher and the researched, even when the focus of the research is on presumably unemotional topics such as computing technology.

Ethnography

Ethnographic work is currently carried out in a range of diverse naturalistic settings by ethnographers from a range of disciplines including anthropology, sociology, nursing, and other professional groups. Many ethnographers examining phenomena within the nursing profession and hospitals are nurses or midwives already working in this setting. Indeed, there has been a call for the blending of nursing and ethnographic methodology, termed "ethnonursing", within the field.[1] The aim of the ethnonursing approach is to formulate analysis which describes the cultural themes that influence the actions of the individuals and the group in the chosen setting, culture or subculture. Leninger[2] reviews some major contributions of the theory, illustrating the importance of being attentive to cultural care diversities and universalities. Specific examples pertinent to hospital life include patients' compared with nurses' perceptions of caring,[3] attachment in a neonatal intensive care unit,[4] and Anglo- and African-American elders in a long-term care setting.[5]

Ethnographic research is concerned with socially negotiated meaning as it is worked out in social settings. As such, primary epistemological importance is given to the accounts of the actors, or members, of those groups. For example, my own research was concerned with the social management of computing artifacts* in nursing settings. Consequently, my thesis highlights

---

* An artifact is some machine or tool of technology. As such it can be applied to either computer hardware, software or paper forms. The term "artifact" has been deliberately used to enforce the ethnographic "strangeness" methodological position, that is, the intent is to force us to look at the familiar in an unfamiliar way.

the way nurses and hospital receptionists talk about and work with computers on a daily basis.

Ethnographic research may draw meaning from a range of information sources including in-depth observations, conversational interviews, organizational records, and other documents. The task of the ethnographer is to reconstruct the meanings of actors within the social group, and the role of the ethnographer is to

> "… participate, overtly or covertly, in people's daily lives for an extended period of time, watching what happens, listening to what is said, asking questions; in fact, collecting whatever data are available to throw light on the issues with which he or she is concerned."[5]

Implicit in this definition is the issue of the reflexive character of social research; that is, the recognition that the ethnographer plays a contributory role in the social world she or he observes. This reflexivity includes the burgeoning study of the role of emotions in the field.[6] This chapter incorporates a recognition of my own reflexivity in this research project.

Background

Prior to this work, my contact with hospitals had been limited largely to weekend visits as a child to see my grandmother in a long-stay geriatric ward in New Zealand. This limited exposure to hospital life left me mostly unprepared for my fieldwork experiences which took place in a variety of care settings. While there were some medical dramas within popular culture (such as television and film) at the time I began this fieldwork,[7] I was largely unaware of them. In addition, I began this fieldwork before the emergence of very graphic representations of medical events, such as those presented in the television program "ER," and before the first televised accounts of live births and deaths in the U.K. Therefore, I was largely a novice or stranger to the hospital setting, a role which had particular emotional and other consequences for me, in both my public and private lives.

Observations

The primary focus of my study was the way in which nurses incorporate and manage computing artifacts on an everyday basis; therefore, I was interested in issues such as how nurses physically attend to information technology, how they talk about computers in the wards, and how they prioritize this work against attendance to patients. In order to gain some understanding of exactly what technology was available and who used it, I determined that it would be worthwhile to undertake some preliminary observations in a local public hospital. With the support of the hospital information technology (IT) manager, nine different areas were selected for

preliminary observations. These areas represented a range of working practices (e.g., acute, general, and administrative areas) in which a variety of
paper-based and computer artifacts were used by administrative and medical staff. These included, for example, nurse rostering systems, patient information systems, and patient support equipment such as patient drips and
electrocardiograph monitors — known as EKG monitors in the U.S. Some
of the systems had been in use for some time in particular areas, others were
used on a trial basis, or had just been implemented. Information obtained
from the preliminary observations was to be used to determine three or more
hospital areas that might form the basis for further in-depth observations
suitable for the primary data collection phase of the fieldwork.

   During the preliminary observations, I spent approximately three days
in each of the areas observing and taking notes about what I considered was
going on. The initial observations were useful in many ways including being
a general introduction to the ward, the staff, computing, and other technology. A range of practical things immediately came to light. For example, I
noted which nurses were trained and had access to the computers, the actual
time of day or night when the computers were being used, the spatial layout
of the ward and the areas where the technology, in particular, was located.
This preliminary knowledge enabled me to make sense of who actually used
the computers — although why and how those particular people did so
required further investigation. The preliminary observations helped me to
further define what was sociologically interesting about the working practices of nurses with computers.[8,9] In particular, I came to believe that the
decision to observe interaction with computers as a form of conversation in
the ethnomethodological tradition within sociology,[10-12] would not do justice
to the ways in which nurses incorporated computing into their working
practices. That is, the two or three minutes in which a nurse might sit at a
computer entering data does not adequately reflect the real investment
involved in the production of this encounter. In a sense, this interaction
represents only the *product or outcome* of the work. This investment was not
confined to the physical entity of the computer, the spatial areas of the ward
or to the time of the actual interaction, as nurses often talked and thought
about this work in other times and places.

   It became apparent that if I wanted to examine nurses' use of computers,
I would have to extend this observation to other artifacts (i.e., to other forms
of technology including low technology tools). For example, it became clear
that in general ward areas, nurses' use of computers is intrinsically linked
with their use of other high- and low-technology forms. By this I mean that
information ultimately entered into computers is discussed, collected, and
made sense of on such low technology tools as paper-based forms, white-
boards, and bedboards. This information is then entered, or copied, into the
computer as data. It would not, therefore, be satisfactory to merely confine
the investigation to computers alone. Rather, attendance to a range of tools
is also relevant to the time, place, location, and content of computer use.
Furthermore, in one area, the accident and emergency department, it was

immediately obvious that *nurses'* use of computers — in terms of hands-on physical contact — was virtually nonexistent. However, it appeared that receptionists' use of, and orientation to, computers in the department was paramount. Therefore, it became useful to extend the observations to the working practices of staff other than nurses. As a result, in order to adequately capture the minutiae of nursing work with technology and other relevant artifacts in the primary observations (which would span a period of some weeks in three of four significant areas), I decided to employ a mixture of methodologies. These might include in-depth observation of working practices, individual and group interviews, and videotape recording of nurse-computer interaction.

Role negotiation: the management of self in the field

The preliminary observations had a range of unintended consequences for the way in which I viewed myself, and how others viewed me, in the field. For example, in some senses, I felt that I did not have a legitimate role to play in the ward. I have no medical training and, unlike nurse ethnographers, I could not contribute, either verbally or otherwise, to much of the nursing work. Even my position as a researcher was unusual in that I was not engaging in auditing hospital records, which is the role of many hospital-based researchers. An ethnographer working in a cancer unit captures this same sense of ambiguity. He says, "… everyone but me seems to be part of a team, a cohesive unit, each member of which has a clearly defined role and is working towards a specific end, namely the treatment and care of patients. I am neither a member of that team nor are our goals the same."[13]

Similarly in many ways, therefore, I did not have a particular contributory function to perform, as my work was not patient centered. Perhaps because I had initially been introduced to nurses in terms of undertaking research concerned with computers, the nurses often cut short any interaction I had with *patients*. Consequently, I found that my role was constantly challenged in the intensive care unit, sometimes called intensive therapy unit, where the patients were seriously ill, while I had a greater level of acceptance in other, more general, areas. Overall, I felt that the level of my participation in hospital life was never clearly defined, even by myself.

Thus, I was largely marginal to the ward life in two ways, first, because I was an outsider to both the ward and the hospital and, second, because I did not have a useful task to perform on the ward. This feeling of marginalization was compounded by my physical inactivity in the ward areas, in contrast with the active nature of nursing work. Ethnographic work in a hospital has one major advantage over many other ethnographic sites: hospitals generally and nursing work specifically are characterized by a paper-oriented culture. Note-taking in this instance is viewed as an everyday, normal activity. This worked to my advantage since I did not have to attempt to conceal my note-taking activity. To the casual observer it might have appeared that I was engaged in normal working practices.

This sense of ambiguity was further reflected in my working practices. For instance, while in some reflexive nursing ethnographies the ethnographer allows members to view and write upon the ethnographer notes, I chose largely not to do so. This was due to practical concerns, such as the concern that the notes might lose their structure, might be misplaced, or that nurses might be influenced by, or try to amend my recordings. Occasionally, individual nurses pressured me into verbally recounting some of my observations and into suggesting preliminary analysis based upon this.* The staff involved usually took these opportunities to correct and clarify particular working practices, and give their view as to how or why particular events happened. These conversations were recorded in my fieldnotes. While I found these encounters to be both mildly embarrassing and stressful at the time, the conversations I had with the staff ultimately proved useful to the research analysis and findings.

Explaining the research

I had some difficulty concerning how best to explain the nature of the research and how much detail to divulge to participants. Thus, I found myself giving different versions of the same truth about my research to different people at different times. Considerations of what to say reflected my understanding of who the person was, what his or her role was, and my perceptions of how sensitive the research might be for that ward or unit (as computer technology was seen as either being successful or a failure in particular areas). This seems to be a common strategy in ethnographic work. So for instance, in McIntosh's[13] ethnography in a cancer ward while, on ethical grounds, he thought it best to tell patients exactly what the research was about, he did not explicitly want to tell them it was about cancer. Although this was a cancer ward the word "cancer" had particular connotations and was actively avoided by staff and patients. Instead he told them the research was primarily concerned with communication. As a result, questions about what they knew, what they had been told, and what they would like to know about their illnesses therefore seemed perfectly natural.

Confusion over my role

It was evident that in each of the areas I observed, both staff and patients also felt some confusion over my role. I had assumed, having informed the ward managers and sisters of what I was doing and having prepared and distributed information sheets for nurses and patients, that an account of what I was doing would filter through to the people on the ward.** This was

* On a couple of occasions senior nurses insisted on viewing the notes. They expressed surprise that I recorded everything that was happening and said in the ward, rather than just noting the detail of the use of technology and other artifacts.
** I distributed information sheets to individual patients in the ward, and posted a general information sheet on the ward notice board. Some patients chose to speak to me about the research, but many did not. Nurses largely rebuffed attempts by patients to talk to me about the research.

a miscalculation. In the wards, the information sheets were usually placed in a drawer and seldom read. I was told this was to prevent them cluttering up the nursing station. Patient and visitor information sheets placed on ward notice boards tended to disappear within a day or two. Hence, I was frequently asked "so what exactly *are* you doing?" by nursing staff. In intensive care, one sister was very cold toward me. When I explained the research to her, she said that it sounded like a "time in motion study." She seemed to think that we were concerned with developing technology with a view to replacing nurses in the unit. Although I went on to explain the research in some detail to her, she still remained suspicious of my intentions and generally made it difficult for me to observe the unit. Ultimately, I decided to withdraw from this area. I later found out that structural changes coincidentally were made to staffing levels in this unit and it might have been that the sister had heard rumors about these changes and was concerned that I was associated with them. Alternatively, she might have viewed my presence as an opportunity to register some form of protest. One major disadvantage for the novice ethnographer is that ignorance of such political machinations might result in the ethnographer being subjected to hostility from those working in the area, without the ethnographer necessarily knowing the reason behind it.

Dressing for the field

In order to avoid confusion about the nature of my role in the hospital, I gave considerable thought to how I should dress in the field. In another ethnographic account of hospital work the researcher had worn a white coat as a way of passing for a member of staff.[14] I chose to dress in smart and plain clothes, with a hospital badge as means of identification. Despite this, confusion still arose concerning my role. In the accident and emergency department, when I stood in the office area, this area might be known as the ER reception area in the U.S., patients often mistook me for a receptionist and became annoyed with my sitting at the desk, and writing notes, when there was a queue of patients waiting to be processed. I told them that I was just observing and that a receptionist would be with them shortly. In the accident and emergency reception rooms, patients often gave me puzzled glances, probably because I did not have any obvious wounds and did not appear to be a member of staff. One patient who had suffered a heart attack asked me if I wanted to see the doctor before him. In addition, I was often approached by members of the public for directions around the hospital. One elderly woman, assuming that I was employed by the hospital, became angry when I could not tell her what the hospital visiting hours were.

In the theater unit, referred to as "surgery" in the U.S., clogs, gown, mask, gloves, and a plastic hat are compulsory dress. Staff are usually identified by the color of their hats. There is no corresponding hat for an observer, so I was given a blue hat — that of a sister. Consequently, I was occasionally asked to perform minor tasks while observing in the operating room.

As I was introduced to the staff by the IT manager, the nurses tended to associate me with the management generally and with the IT staff in particular. This confusion was further compounded by my hospital ID badge which read "researcher, IT department". It was apparent that many staff members were concerned that I was there to assess their nursing knowledge or skills. I repeatedly used my lack of nursing background to reassure them that I could not legitimately do so — an option obviously not open to nurse ethnographers.

In most areas, the nursing staff considered that because I was researching technology I must be an expert with computers. Consequently, in all the areas and on more than one occasion, I was called away from taking research notes to attend to computer and printer problems which generally I was unable to resolve. Another consequence of this assumption was that staff assumed that I wanted to view the computers only when in use; although I was equally as interested in times when they were not being used. In addition, they often apologized for not waiting until I was present to use the computers, or they offered to repeat what they had done earlier, while I watched. In one instance when I commented on information recorded on a paper form which had been fed into the computer, the nurse interpreted my comment as a criticism and offered to reinput the information. Further, during one videotaped recording, a nurse asked me if I could rewind the videotape so she could start again as she had made a mistake. Thus, initially at least, there was a concern on the part of nurses that as I was an expert with computers and that I only view use of technology at its best. In these instances, I sought to emphasize that I wished to view naturally occurring use — warts and all.

Largely to combat feelings of marginality, I actively sought to participate in the everyday ward life in small ways. After a few days in each area, I either volunteered or staff would ask me to perform small tasks. In the ward areas I would keep an eye on a patient to ensure that he did not pull out his drip or fall out of bed. In orthopedics, I would tuck the blankets around a confused patient, fetch a vase for flowers, or make a cup of tea for the nurses. In the Accident and Emergency department, I would help to do some filing or tell waiting patients that the nurses would be returning soon. In the coronary care unit, I helped nurses with their own research projects. However, I found that ethnographers in hospitals have to consider carefully how much they can participate in ward life without jeopardizing their separateness. This is particularly relevant when dealing with potentially litigous situations. For example, after observing an operation in the theater unit, a surgeon asked me to help lift an unconscious patient onto a gurney. I was concerned that he thought I was a nurse and I pointed out that I was only observing. He replied sarcastically, "Well, you can lift can't you?" I felt compelled to help although I was concerned about the legal ramifications in case we dropped the patient, or if I had injured myself while incorrectly lifting the patient. Notably, surgeons are not trained in lifting techniques, and do not do so routinely, even in the theatre unit, other staff are usually on hand who have been trained in lifting techniques. Perhaps the surgeon was also saying something about the usefulness, or otherwise, of nonmedical researchers in this arena.

These are considerations that have to be worked through during the course of the fieldwork. Each encounter has to be judged on its own merits and managed by the ethnographer. Further, the ethnographer must consider and reconsider to what degree she can immerse herself in a particular culture while still maintaining some sense of *strangeness* from that culture, and while still maintaining her nonparticipant role. Ethnographers may have to consciously work to avoid the pitfalls of "going native."[15] See for example the difficulties encountered by a researcher observing female police officers at work who "entered their world and lost sight of... (her) own."[16] This dilemma became particularly apparent to me in one of my observations in the accident and emergency department.

Dilemmas in the field

During this particular observation, the office area was very busy, there were a number of patients waiting, and there were no receptionists present. A clerical worker who was employed to work on a separate research study from mine and who occasionally helped out when there were a number of patients waiting, used the computer to take patients bookings and forward them to the waiting room. I observed the clerical worker booking in a child with asthma who had appeared at the window of the office which was generally used by patients with minor injuries. She then sent the child and his mother to the minor injuries waiting room.* From my previous observations in this department I knew this wasn't normal procedure and asthma attacks — particularly where they pertain to children — are considered life threatening. I was unsure if the clerk was aware of this. I spent some time deliberating with myself about what to do. As an asthmatic myself, I knew that asthma attacks could be frightening and I was concerned that the child might stop breathing, have a heart attack, or die while waiting in the "wrong" waiting room. However I was concerned that if I brought the matter to the attention of the clerk, I would be working outside the boundaries of my self-imposed role by appearing to question her decision-making about how best to deal with this case. Further, I was concerned with disrupting what I considered to be the natural flow of events. After a few minutes, a receptionist with whom I had developed a good rapport, returned to the

---

* Patients sent to the minor injuries waiting room undergo a process of triage by a nurse and are seen in order of the seriousness of their injuries or complaint. Typical examples of minor injuries include cuts, abrasions, suspected broken bones, etc. Patients in the minor injuries waiting room may have to wait a number of hours before they are seen by a physician. Those patients with major injuries or a serious illness may see a physician straight away as it is assumed their injuries might be life threatening. Typical examples of major conditions include chest pain, suspected drug overdoses, and asthma. Particular extenuating circumstances may alter the classification of the patient; for example, if the patient is considered to be drunk.

office. I was still concerned for the child and decided to mention the matter to her, away from the hatch area. As a result, the receptionist was concerned and went to speak to the clerk. The clerk told the receptionist she had already spoken to the triage nurse directly about this patient, and the triage nurse had said the patient should wait in the minor end. The receptionist relayed this information back to me.

This encounter was problematic because I felt that I had damaged my relationship with the clerk by stepping outside the boundaries of the non-participant role I had established for myself. The clerk was particularly cold toward me for the rest of the shift, and I felt so uncomfortable that I decided to end this observation earlier than I would have normally done. The following is what I recorded in my fieldnotes at the time:

> I feel a conflict of interest here — how should I behave in delicate situations like this? How much should I interfere in the running of the A and E office life... I know, from my experiences with asthma, that it can be life threatening, and I realize from my observations that the receptionists routinely put asthmatics through to the Doctor in the major end. The boy looked pale and was coughing, crying and struggling for breath. His mother looked anxious and was clicking her tongue. I felt guilty for drawing this case to J---'s (the receptionist) attention but I was concerned that if I didn't, it might make things worse for D--- (the clerk) and/or the boy.

This situation made me aware of a number of things. First, that I had developed an idea of what was and what was not "normal" behavior for staff; second, that I had certain biases where staff were concerned (e.g., I considered that some staff were more aware of these normative rules than others); and third, that I was not — and could never be — totally objective. My interpretation of the seriousness of the situation was based on my personal experiences and my understanding of the normative rules of that area. Ethnographic researchers might often encounter situations in which their personal feelings are at odds with their professional role. These may be of particular concern for the hospital ethnographer who is located in an arena where illness and sudden death are commonplace.

A sense of belonging

Many of these issues feature in nonparticipant research in a range of arenas, but some issues are specific to research in hospitals. Particular problems arose for me during the preliminary observations, as I had to renegotiate entry into nine different areas and try to establish rapport with a range of people — doctors, surgeons, nurses, administrators, and patients — in each of those areas. During the preliminary observations, the short time I spent in each area, an average of three days, compounded this problem. In particular, it was very difficult to develop any rapport in theater (surgery) where there are many teams of staff working simultaneously, and where I found it difficult to identify people as the staff all wore the same gown and often wore masks. Notably however, in the theater recovery area, I gained a deeper sense of rapport with the nursing staff as they were unmasked, were fewer in number, and remained constant to this area.

During the primary stage of the fieldwork, my observations extended over a period of eight weeks or more in each area which gave me more time to come to know the staff and gain their confidence. In one area, the coronary care ward, I developed a particularly close relationship with the nurses, perhaps because they were a small group of nurses who had worked closely together for some time. One of the nurses, in particular, appeared to have taken me under her wing. I felt that we had an affinity because of the technological character of this unit which fascinated me. Indeed, I felt able to discuss things with the nurses in this area which I would not have felt comfortable doing elsewhere. However, this level of closeness also caused difficulties, as will be discussed later in this chapter.

Practical considerations

For every ethnographer, there are a range of practical considerations that have to be taken into account when beginning observational work. These issues include such things as where the ethnographer and her equipment may be located; whether she can freely move around, and if so in what areas; and with whom she may talk. The importance of these factors is that they may shape both the relationship between the researcher and the researched and the content of the fieldnotes.

Uncharted territory

When I first began the observations in the hospital, I experienced difficulties associated with moving around the hospital generally and with not knowing how the hospital systems operated specifically. These difficulties included things as minor and as embarrassing as walking into an unmarked male changing room in the theater unit, not realizing that the "sluice room" was the room in which the nurses disposed of patients' bodily wastes, and entering a sterile area in plain clothes. These problems became fewer as I spent more time in the field, and it became apparent that there were certain boundaries that existed across different wards and units concerning where I could go, and how I must behave. For example, while I could enter some semiprivate spaces such as nursing stations and, on occasion, sisters' offices, I would not be allowed access to private spaces such as behind curtained bed areas or in isolation rooms without an accompanying staff member.

Negotiating a space

Other issues involved the physical placement of myself and my belongings on the ward. The senior staff, perhaps considering that I was primarily interested in computer usage (and perhaps with a view to also observe me at work) placed me near the computing equipment — usually in the nursing station area. Consequently, I had less choice about my placement in the ward areas than in the accident and emergency department, where computers were placed at either end of the office. In the wards, I could observe what occurred in the nursing station and down the ward to some extent, although my view into the actual bays was restricted. In the accident and emergency department, I tended to position myself inside the office area, where I could observe the arrival of patients and follow their progress to either the major or minor waiting rooms. My location during the observations naturally meant that I observed some events more than others. For example, although there were computers other than those in the accident and emergency department office, I seldom saw these in use and had to ask permission to do so. This meant that my work in this area focused on what the receptionists did while in the accident and emergency office. I did not view, and therefore largely cannot account for, other aspects of work which may have some bearing on the way in which patients are attended to and categorized in the department. For example, other researchers have addressed the categorization of patients by doctors, ambulance, and other staff.[17-19]

Obviously it is impossible for the ethnographer to see everything at once; however, we have to consider how the ethnographers' view, both visual and cognitive, constitutes the subject which is being observed. That is, because I located myself, and was located, in areas where computers and monitors were physically prominent and where nurses tended to paperwork, the use of particular artifacts were prominently featured in my observations and analysis. For instance, I discuss in some detail the use of particular computer systems such as NMS (nurse management system), ANSOS (a nurse rostering system), ECG (electrocardiograph) heart monitors (known as EKG monitors in the U.S.), while other equipment such as pulse monitors and ventilators are seldom mentioned in my fieldnotes. Furthermore, I had to make a conscious effort to record the use of other artifacts such as telephones, which were very familiar and mundane and may, therefore, have become "invisible" to my eye.[20] I consciously sought to record when the telephone rang, who answered it, and the nature of the call, where this was discernible. Many other routine events, one example being the automatic monitoring of a patient's blood pressure, were often missed as these events became less strange to me and my attention was drawn elsewhere. During the major areas of observation, I deliberately structured my fieldwork to try to maintain some degree of strangeness. For example after spending a week observing in coronary care, I would leave the field for a week and engage in other activities (for example, reviewing my notes, reading the literature, etc.), before returning to the field.

Placement of my equipment and belongings was also problematic. There are no obvious places to hang a coat or place my bag in the wards. I thought it significant that nobody suggested I store my belongings in the staff cupboard, although it was apparent that they hindered staff in their work. In cramped quarters, nurses had to ask me to move in order to remove drug trolleys, for instance, or to gain access to filing cabinets. In orthopaedics, there was a cardiac arrest and a nurse could not access the phone quickly because I was sitting at the nursing station writing notes. It seemed to me that not inviting me to use the community storage areas was one way of reinforcing my role of stranger or outsider.

On the move

In some areas, such as casualty, I was able to move around the office and waiting rooms without attracting undue attention — perhaps this is because there were other groups of individuals often wearing plain clothes such as porters, ambulance personnel, members of the public etc., who were also doing this. In other areas, such as ward areas, staff appeared confused and asked if I "wanted help" or if "I was leaving" when I got up from my seat in the nursing station. I felt very frustrated by these restrictions, while understanding the nurses' concerns with protecting the privacy of their patients. Consequently, I had to find other, more legitimate ways of moving around the ward. For example, I sought permission to accompany nurses during the "handover" period when they prepare for shift changes by discussing each patient's condition and their care at each patient's bedside.

In theater, a senior nurse arranged for me to follow individual patients' progress through the theater system, thus dividing my time between three major areas of interest to me: the computers at reception, the computers in a particular operating theater and those in the nursing station in the recovery area. In theater, I tended to leave a particular area when the patient I was escorting also left that area. This proved difficult when the nurses ran alongside the trolley which carried the unconscious patient from the operating room to recovery. I felt quite conspicuous running alongside when I had no legitimate reason for doing so, and when I was unsure why this was necessary.* At the same time, I thought that to walk between the areas or to stay behind would draw unnecessary attention to myself. This is surprising given that I did not attempt to imitate the actions of staff in other areas.

Some of these problems might be specific to hospital settings. There seem to be distinct divisions of space within wards: public spaces (open ward areas, reception areas, corridors), semi-private spaces (in nursing stations, behind receptionists desks, in kitchens), and private spaces (sluice rooms, behind closed curtains around patient's beds, sisters' offices, isolation

* Initially, I assumed that staff ran with the patient from the theater room to recovery because they were concerned the patients condition might deteriorate during the brief trip down the hall. However, I was told by a nurse that staff run because they are very busy and have a tight schedule of operations to attend to.

rooms). My uncertainty about others' perceptions of the limits of my legitimate access to these areas meant that I was careful about which of these
spaces I entered. I was usually very reticent to enter private spaces and either
asked or waited to be asked before I did so, as the following fieldnotes imply:

> There do not seem to be any nurses on the ward. Per-
> haps they are in the office down the hall doing the
> change-over? I debate with myself as to whether I
> should join them. Would this be acceptable? What ex-
> cuse could I give for interrupting them? T--- (senior
> nurse) walks past. Says to HD "Do you want to see the
> handover? It's down the hall, second on the left". I get
> up and go to the office, bringing a chair with me.

These differing levels of patient contact depend upon the particulars of
each area: the size of the ward, the number of staff on duty, the severity of
patients' conditions, as well as the length of time the patient and I spent in
that particular ward.

Misreading the signs

I also experienced some difficulty with the terminology which most of the
staff took for granted. Consequently, I found it difficult to note down exactly
what was being said because I was unfamiliar with the terms being used.
This was particularly evident in highly technological areas such as intensive
care, theater and coronary care. Even common terms were sometimes mis-
interpreted in my fieldnotes, as the following example implies.

The receptionists in A and E referred to two opposite, open ends of the
A and E office area as the minor and major hatches — the distinction relating
to the classification of the patient's injuries. Briefly, a patient with a minor
injury would be triaged by a nurse before seeing a doctor, a patient with a
major and potentially life-threatening injury would see a doctor virtually
immediately. This created some initial confusion for me as the following
fieldnotes indicate:

> It has taken me some time to understand the recep-
> tionists' distinction between the minor and major
> hatches. For some time I assumed that minor referred
> to patients under a certain age and major referred to
> patients over a particular age. This idea was probably
> formed by noticing that mainly elderly patients ap-
> proached the major hatch and young people and chil-
> dren tended to approach the minor hatch.

When I read earlier fieldnotes, I often pick up on these mistakes and
things which I noted down at the time but which were unclear to me have

now become clearer; however, my inability to understand exactly what was being said had both advantages and disadvantages. For example, when I missed the detail of what was being said in technologically intensive areas, I had to ask the nurses to repeat what had been said and explain the meaning of particular terms. Occasionally these explanations elicited more information than I had previously expected. For example in explaining what she meant by a particular term, a nurse had to formulate an explanation in her mind for her working practices and draw upon other artifacts in the process. She might draw diagrams using examples from patient notes or refer to various monitors. Thus, the explanation provided by this nurse incorporating artifact usage could then become the focus of further analysis and discussion.

Nurse ethnographers might not have these sorts of difficulties since they would already be acclimated to the nursing culture and its terminology. However, there is a danger that nurse researchers might not adequately maintain an intellectual distance in their research, accepting the taken-for-granted terminology and work practices of participants uncritically. For example, Landau had to confront this problem:[16]

> I had ideas about what good nursing is as opposed to bad and ... very definite ideas about what is right and wrong, ethical and unethical nursing practice ... how could I possibly presume to be an unbiased researcher? Thankfully, I didn't. I knew I was biased and had to learn how to observe, understand and not make judgments. My questions constantly changed and I learned that I was not in the field to "prove" anything.

The importance of flexibility

As I have argued, hospital ethnography offers many restrictions on the ethnographer's movements. Additional complications arise in that wards and units are subject to continual change because of perceived shortages of beds and staff and changes in patients' conditions. Thus, patients are relocated to other wards and other hospitals as their condition changes, and as other more urgent cases, require nursing care. This was particularly the case in small, specialist units such as coronary care.

Frequently, when the coronary care unit was quiet, it was closed down and the nurses were transferred to other wards to help out. When the unit was busy, additional staff might be called in to help out and mundane tasks, such as attending to the computers, would be left until things were quiet on the unit and there was little to do — primarily, in the very early hours of the morning. The changing nature of work on the wards ensured that I had to be very flexible and adaptable. While in some areas such as accident and emergency I could observe computers being routinely used on an ongoing basis, in other areas such as orthopaedics, computers were generally only attended to on the night shift — usually around two or three a.m. This meant that I had to be flexible and prepared to stay up all night, travel some distance from my home to the hospital, or stay in hospital accommodation, in order to observe and videotape this interaction.

In accident and emergency, I thought it appropriate to observe specific times of the week when the patients and their injuries were different from other times of the week. For example, it was well known within the department that patients suffering from drug overdoses were likely to arrive on Friday and Saturday nights, car crash (also known as "RTA" or road traffic accident) victims arrived late in the evenings, and a large number of sporting injuries occur on Saturdays and Sundays.

Good rapport with staff was essential so that I was given permission to observe at these times and so that staff would contact me when sudden, unforeseen changes occurred in the wards. If a good rapport was not developed, as was the case in the intensive care unit, staff would simply "forget" to let me know when they would be performing particular tasks such as entering patient details into the computer, connecting patients to ECG monitors, or updating the nurse roster. This is particularly problematic when the ethnographer does not live near the hospital, has heavy equipment to carry, or has specific procedures she wishes to observe.

Emotional considerations — the management of emotions

Little of the ethnographic literature addresses circumstances in which the ethnographer encounters emotionally taxing situations and may have to consciously manage their, and others, emotional responses. Ethnographic work in hospitals may be particularly challenging and calls for a conscious management of emotions by the ethnographer as he or she encounters scenes of illness and sudden death.

The ambiguity of the nonparticipant role became most apparent to me when, consciously or unconsciously, my emotions became engaged during my fieldwork. Nurse ethnographers might know what can be expected emotionally in a hospital setting. They might not have to consciously consider how to act. For the non-nurse, the intense personal commitment and the large amounts of time spent in the field called for by ethnography make this an important area for consideration. My own academic preparation for the field was more concerned with who, what, and how to observe, rather than the emotional consequences of that observation for both observer and observed. This lack of preparation in terms of academic training may prove very difficult for particularly young or inexperienced ethnographers, or those working in particularly acute or threatening areas such as psychiatric units.

Goffman[21] notes the normal requirements of tact and courtesy and "inter-action ritual" in general mean that part of the ethnographer's role in the field is ongoing, self-conscious impression management. While in the field, I encountered a number of upsetting incidents in which I felt that I had to consciously manage my emotions as a means of coping with the situation.

Often, the first tactic I employed in these situations was to look to the nursing staff for clues as to what would be appropriate behavior on my part.

Culture shock

The process of estrangement first experienced by researchers in the field is frequently called "culture shock." My first observation took place in the intensive care unit. This was not by choice. There had been no opportunity to begin the observations in general ward areas and scale up to acute areas. Rather, the order of the preliminary observations had been arranged to best coincide with other events, such as the installation of technology in the wards and units. I can still recall my feelings when I first encountered the ward. The first impression that assailed me was the smell of antiseptic. The patients were lying naked, covered only by a sheet, with various tubes and machines sprouting from their bodies. They were indistinguishable by sex or personality, because they were either in a coma or unconscious, and looked dead. One patient was black and blue with bruising. Only the machines seemed to be alive; they bleeped constantly. The nurses were dressed in green overalls and clogs, similar to those used in operating theaters. To my mind, the scene resembled something out of a science fiction movie. This experience was far removed from my pleasant childhood memories of visiting my grandmother in a brightly lit, warm, and colorful long-stay ward in New Zealand (although this often provided a topic of conversation with elderly patients in hospital). It is possible, therefore, that the hospital ethnographer might experience a sense of estrangement, not just upon first coming into contact with a hospital, but also when entering different medical settings within the hospital or entering a similar hospital setting in a different country. For example, one experienced medical sociologist and ethnographer seasoned in work in teaching hospitals in the U.K. recounts many unfamiliar features of life and medicine in an American hospital setting.[22]

Emotional situations in the field

In one observation, early on in my fieldwork in orthopaedics, a female patient, located about six feet from me was moaning loudly, as if in pain, and calling out "Oh, dear, oh dear, what am I going to do?" The nursing staff did not appear upset or disturbed by this patient's behavior. They did not seek to comfort her and rarely talked to her, even in passing. I was located across from this patient for three consecutive days. I found the patient's behavior very upsetting. I would leave at the end of each observation feeling exhausted from trying to maintain a facade of indifference to her distress. This is what I wrote at the time:

> I feel uncomfortable with the patient opposite me
> moaning — particularly when it becomes quite loud
> and distinguishable words can be heard. I feel as

> though I should comfort her in some way — pat her
> hand, tell her everything is okay — something. But I'm
> not sure if this would be acceptable to the nurses or
> the patient — it might make things worse. So I do
> nothing.

Other instances I found particularly trying were when women came into
the accident and emergency department in a distressed state, apparently
suffering from a threatened or real miscarriage. Around this time, I too had
a miscarriage, an experience which I had found surprisingly difficult to deal
with emotionally. This experience had been compounded by the fact that I
felt that my local general practitioner had adopted a largely unsympathetic
approach toward the situation, and I had not felt able to talk to anyone about
the experience as I did not want family, friends or work colleagues to learn
that we had been trying for a baby. It is perhaps significant that the first time
I felt able to discuss this experience is when the conversation arose in the
context of discussing the distressing nature of nursing work with seriously
ill patients in the coronary care unit — roughly one patient dies each week
in this small unit, which caters for a maximum of five patients at a time. The
nurses were very sympathetic toward my experience, and I believe that
sharing this personal tragedy with them enabled them to feel more comfortable with me when discussing particularly distressing events that had
occurred within the unit.

In these and other circumstances, I found myself actively struggling to
manage my emotions. During a routine observation, I casually observed an
elderly patient and his family who were seated directly opposite me at the
nursing station. Without any warning the patient simply stopped breathing,
and did not breathe again. A young female relative cried out. It took me a
couple of seconds to realize that the patient had died. Although his family
were gathered around him and a nurse had told me the patient was "poorly,"
I had not realized this meant that he was close to death. This was another
instance where a nurse ethnographer might have had an advantage. The
nurse ethnographer might have been able to deduce from the patient's
appearance, the terminology of the nurse, and the positioning of the patient
next to the nursing station, that the patient was near death. A nurse, hearing
the woman's cry went to the bedside and said, "Oh, he's gone then," and
pulled the curtains around the patient before simply turning and walking
away. I was slightly shocked at what appeared to be an uncaring attitude
from a professional engaged in what I perceived as "care-work." I further
felt very upset because I had not seen anyone die before. After a couple of
minutes, the relatives, visibly upset and crying, left the ward walking past
me as they went. I had to consciously fight back my own tears and quell a
desire to leave. I was concerned that the nurses would realize I was not the
passive observer they believed I was. It took me some time before I felt
capable of making my good-byes to the nurses on the ward. After leaving
the hospital, I felt haunted by this incident for some days and vaguely

annoyed at myself for feeling that way, when I had not even met the patient. I also felt guilty that I had not said anything to an elderly male patient lying in the bed next to the dying patient. He appeared to be particularly upset at the death of the other patient, and I wondered if they had become close during their stay in the hospital. At my interview, I had been assured that should something upsetting occur during my fieldwork, there would be a member of staff whom I could call upon to talk things over with (although this person was never named). However, even when not physically located in the hospital, I was still concerned about jeopardizing my "competent researcher role," particularly with my colleagues at the university. That is, I consciously sought to reaffirm my image as a distanced observer who was largely uninterested in the lives of patients except where they pertain to the technology in the ward. However, despite a desire to achieve some sense of impartiality, I recognize that ethnographers, like any researcher, can never be truly "neutral."[23] Consequently I largely kept these feelings and experiences to myself.

On some occasions, I consciously, or otherwise, acted outside the boundaries of my impartial observer role. This proved to be to the detriment of other people with whom I came into contact. In one encounter in the Accident and Emergency department, a mother and child approached the hatch area. At this time I was occupied, writing in my fieldnotes book. I heard the mother inform a receptionist that her three-year-old son had rubbed chemicals in his eyes. I looked up from my fieldnotes, straight into the gaze of the boy, whose eyes were red and swollen and cheeks were tear-stained. I unexpectedly flinched at the sight. The young boy, who had been calm until then, saw my reaction, and his face crumbled. He started to cry loudly for a long period of time. I felt terribly guilty and ashamed that I had further upset him and ducked my head, trying to appear busy with my work. Not only did the noise of the crying child clearly upset others in the immediate vicinity, it also impeded the booking in process as the receptionist had to ask the mother the same questions a number of times in order to be heard over the noise. After this encounter, I always attempted to maintain a blank expression when faced with such sights and I felt a sense of satisfaction when I had managed this in difficult circumstances. I found it surprising that, while I could quite comfortably spend hours observing surgical operations at close quarters, with no incumbent feelings other than that of fascination, it was usually the mundane which disturbed or upset me most. I can only assume this was because I was able to distance myself more from the "strange" than from the mundane events.

The reactions of the hospital ethnographer to upsetting incidences do not always have a negative effect upon the emotions of others. For example, in one particular encounter involving the unexpected death of a patient on night shift in an orthopaedics ward, the nurses were visibly upset, and the senior nurse in charge was close to tears. She sought reassurance from me that there was nothing that she could have done to save the patient. I found myself rubbing her arm and reassuring her that this was so, although I have no medical training at all and I did not know if this was the case. After the senior nurse had spoken to me in this manner, the other nurses continued to seek reassurance from me during the next ten minutes or so. I think this is interesting in terms of the nonparticipant role. An ethnographer, with an ambivalent role, can find that her role enables participants to do things that they might not normally have done had she not been there (although we have no way of actually assessing this). For example, it is difficult to imagine a senior nurse crying or seeking reassurance from others in her vicinity, such as her junior nurses or other patients, had I not been present at this time.

It also became apparent to me how much the death of a patient can affect others in the ward. I noted in my fieldnotes that the nurses talked differently to the patients that morning. They tended to be less patient with them, and the senior nurse became angry with a patient when he gripped her arm too tightly and would not let go. Furthermore, the atmosphere in the ward seemed particularly tense, some of the patients informed others who were waking about what had happened, and a noisy argument broke out between two of the patients, which the senior nurse had to attempt to resolve.

This particular example illustrated to me that the actions of staff, patients, and ethnographer also need to be considered and understood in the light of emotional circumstances. Hammersley and Atkinson[5] argue that

> "personal reactions to clinical encounters — fascina-
> tion, revulsion, and embarrassment for example —
> cannot simply be used to extrapolate to the feelings of
> others such as doctors and medical students. However,
> they can be used to alert us to possible issues such as
> the process… whereby medical practitioners' most ex-
> treme feelings may be masked or neutralized."

This encounter helped to reassure me that nurses also have to con-sciously manage their emotions when dealing with circumstances like this, and it enabled me to feel a closer affinity to the nursing staff. In the light of this analysis, we might reconsider the earlier example of the "uncaring" nurse toward the patient who had died. What might have been viewed as an uncaring act of closing the curtain around the patient, might have, in fact, been the act of a caring practitioner. That is, she provided the relatives with some privacy to express their emotions, and she protected other patients from seeing both the grief of the family, and the body of the dead patient. Her comment "Oh, he's gone then" might further indicate that this death was not unexpected. This might, therefore, have been reassuring to other patients nearby. While we cannot determine the feelings of the nurse in this instance, it might further be that she was as upset as I was. Like myself, she may have sought to disguise these feelings behind a mask of professionalism so as not to upset others around her.

A particular strategy I found helpful in dealing with these feelings was to try to use what I considered to be legitimate outlets for my emotions. I found that recording details of these encounters in my fieldnotes and talking about my experiences in university seminars gave me some relief from these feelings. This is a common strategy within qualitative research.[16] Feedback from other researchers engaging in ethnographic work in other arenas, such as that in children's homes,[24] reassured me of the appropriateness of these emotions and my reactions to them. While traditional ethnographic accounts have tended to exclude analysis of the role of emotions in fieldwork, I came to believe that my emotions and those of others around me played a significant role in the interactions between staff and patients (and by extension the use of technologies). Consequently, these ideas and themes became incorporated into my analysis — particularly in discussion of those areas such as accident and emergency and coronary care, where it was apparent to me that staff had to actively manage their own emotions and those of their patients in order to attend to their everyday working practices. Indeed, in some situations I have argued the active management of emotions by nursing and other staff is central to the management and use of particular technologies such as ECG monitors.[25]

This following is a final example based in the coronary care ward. When patients' hearts are engaged in an unusual aberrant rhythm called atrial-fibrillation, a procedure called an elective cardioversion is performed routinely. The patient is given an anaesthetic to ensure she is unconscious, and then an electrical current is applied to the patient's chest through rubber conductors to shock the heart back into a regular rhythm. By my seventh observation in coronary care, my final observation area, I had developed a particularly good rapport with staff, to a point where they often overrode the wishes of others (e.g., doctors, X-ray staff) in order for me to observe particular procedures. I was particularly interested in observing an elective cardioversion as I knew they were performed routinely on the ward and I had not had the opportunity to see this procedure before.

During this observation, a number of staff moved into a small side-room. One nurse positioned herself next to me and started a conversation about the noisiness of the alarms, while I frantically attempted to note down everything that was happening during the complicated procedure which involved an anaesthetist, two doctors, and two nurses. Feeling distinctly harassed but not wishing to appear rude, I just nodded my head in reply. Finally she appeared to notice I wasn't responding to her. She then asked me, "Doesn't this all worry you?" I replied, "No" — and that I found it very interesting. She said that she used to find it very upsetting and that she had often burst into tears imagining that it was her on the bed. She said she had had a heart attack in the past and had been admitted to a coronary care unit in another hospital. I finally realized that she had been talking to me to try and take my attention away from the events in the room so I would not become upset. I told her that I had seen some upsetting things in the hospital but that this did not worry me. She seemed reassured and said that she could not bear to see anything more. She left the room. I felt relieved that I could concentrate on note taking again.

This incident reassured me in a number of ways. I recognized that I had become accepted as a member — albeit one with a slightly different agenda — in the coronary care unit. The nurse's attempt to protect me, while vaguely annoying in that I wanted to concentrate on taking fieldnotes, was also endearing. It further reassured me that I was not alone in finding even some familiar routine procedures occasionally upsetting. I felt slightly guilty however that I did not attempt to comfort the nurse because I was "too busy" doing research work. Perhaps nurses also feel guilty for not finding the time to comfort patients and relatives who are upset. This is an issue which has recently been the subject of some discussion within the nursing literature.* See for example Reference 27.

A reconsideration of roles?

Emotional labor of this kind raises questions about the role of nonparticipant observer: how can the ethnographer be said to be a nonparticipant when she is consciously, and sometimes involuntarily, engaging in emotional work? While there have been some attempts to map out roles in the past,[28,29] these are fairly limited in their analysis. Gold's typology extends from the complete participant, where the researcher is totally immersed in a culture, to the complete observer, where the researcher has no contact at all with those she is observing. The "observer as participant" and "participant as observer" roles fall somewhere in between and are largely limited to community studies and one-visit interviews, respectively. Within this typology my work would fall somewhere in between these "observer" roles. However, as I have illustrated, the role which I played, perhaps best termed a "nonparticipant observer," is itself problematic given its complexities and the variety of situations in which I, consciously and unconsciously, had to manage it.

Therefore, while not necessarily advocating a reconceptualization of the theory of social roles in fieldwork — which would be difficult as we would have to recognize the constantly changing nature of field roles — I would argue that it *is* useful to constantly revisit ethnographic data, both during and after the observations have been carried out, to look back on ethnographer-nurse encounters, and analyze them as data in themselves. The recording of encounters like those mentioned previously in this paper, are also useful in terms of retracing analytic themes as they emerge in the data, and addressing how the ethnographer's interactions with members of the community/field may have shaped ongoing interaction. Upon revisiting my early fieldnotes, I was surprised to find a number of occasions where I had made unconscious prompts to nurses as they approached the nursing station, asking them when and if they were going to attend to particular pieces of paperwork or computer technology. What is particularly interesting is that

* Indeed, the term "caring" brings to mind images of nurses spending time at the patient's bedside talking to (and where appropriate) touching their patients. This is an image, which much of the nursing literature commends and laments the loss of. See for example Reference 26.

in no case did these prompts result in the unconsciously desired action; that is, although the nurse gave some indication of when she might attend to that work, she carried on with whatever she was doing at that time. The nurses simply did not attend to this work at that time because they did not view it as being appropriate to do. Notably, unsolicited encounters between ethnographer and nurses, such as those outlined above, were more likely to culminate in a conversation, an interaction, or an emotional response, which proved useful for the purposes of the research.

Conclusion

In this chapter, I have explored some practical considerations and emotional consequences of engaging in ethnographic observations in a public hospital setting. These issues are not independent of each other. That is, when engaged in ethnography, practical considerations — such as how to gain access to anticipated areas of interest, and decisions about how to dress, where to sit, what and whom to observe, in what order to observe things, and how long to spend observing in each area — will have a range of implications for the researcher and researched. These implications might include what will and what will *not* be recorded by the ethnographer, and how the ethnographer views herself and is viewed by others in the field. Furthermore, the wide range of people the hospital ethnographer encounters in the field (which may include nursing staff, medics, patients, and relatives of patients), and the amount of time spent in the field often require the researcher to engage in emotional labor of some kind. This emotional labor might range from the ethnographer's unconscious reaction to the emotions of others in the field and the implications of this for the research, to conscious considerations on the part of the researcher concerning how best to manage her own emotions in the field. These are particularly pertinent problems for the non-nurse ethnographer who may be unfamiliar with the medical arena. The examples cited in this chapter have highlighted the ambiguities of a so-called nonparticipant role for the ethnographic researcher. This is a role which may benefit both the researcher and the researched as it allows for more flexibility in behavior than is usual for the traditional nurse-ethnographer or other insider in the field. These are issues which would benefit from further investigation.

Ultimately, we need to recognize the constant interplay between the personal, the emotional, and the intellectual work of the ethnographer in the field. This is particularly pertinent in the field of health and illness where ethnographers may encounter situations of illness and sudden death. Furthermore, this chapter implies that even when research is concerned with topics that are traditionally viewed as unemotional — as in the study of interaction with inanimate objects such as computer technology — an emotional dimension may still exist. The recognition, recording and exploration of this emotional dimension within traditionally nonemotional subjects would benefit researchers and researched alike.

Acknowledgment

I would like to thank Dr. Allison Kirkman at Victoria University of Wellington, New Zealand, for her comments on this chapter. The research reported here was undertaken on behalf of the University of Luton, England.

References

1. Leninger, M., Ethnography and ethnonursing: Models and modes of qualitative data analysis, in *Qualitative Methods in Nursing,* Leininger, M., Ed., Grune & Stratton, Orlando, FL, 1985, 33–71.
2. Leininger M., Culture care theory, research, and practice, *Nursing Science Quarterly,* 9:(2) 71–78, 1996.
3. Patistea E. and Siamanta H., A literature review of patients' compared with nurses' perceptions of caring: Implications for practice and research, *Journal of Professional Nursing,* 15, 302–312, 1999.
5. Hammersley, M. and Atkinson, P., *Ethnography: Principles in Practice,* Routledge, London, 1983.
6. Rosenberg, M., Reflexivity and Emotions, *Social Psychology Quarterly,* 53, 1, 3–12, 1990.
7. Karpf, A., *Doctoring the Media: The Reporting of Health and Medicine,* Routledge, London, 1988.
10. Greatbach, D., How do desk-top computers affect the doctor-patient interaction?, *Family Practice,* 12, 32–36, 1995.
11. Greatbach, D., Conversation analysis: Human computer interaction and the general practice consultation, in *Perspectives on HCI: Diverse Approaches,* Monk, A. and Gilbert, G., Eds., Academic Press, London, 1995, 199–222.
12. Heath, C., *Body Movement and Speech in Medical Interaction,* Cambridge University Press, Cambridge, 1996.
13. McIntosh, J. R. B., Processes of communication, information seeking, and control associated with malignant disease in a hospital ward, Ph.D. dissertation, Aberdeen, Scotland, 1975.
14. Hughes, D., Lay assessment of clinical seriousness: Practical decision-making by non-medical staff in a hospital casualty department, Ph.D. thesis, University of Wales, Swansea, 1980.
15. Lofland, J. and Lofland, L. H., *Analyzing Social Settings: A Guide to Qualitative Observation and Analysis,* 2nd ed., Wadsworth Pub. Co., Belmont, CA, 1984.
16. Ely, M., *Doing Qualitative Research: Circles within Circles,* Falmer Press, London, 1991.
17. Jeffrey, R., Normal rubbish: Deviant patients in casualty departments, *Sociology of Health and Illness,* 1, 90–107, 1979.
18. Godse, A. H., The attitudes of casualty staff and ambulance men toward patients who take drug overdoses, *Social Science and Medicine,* 12, 341–346, 1978.
19. Dingwall, R. and Murray, T., Categorization in accident departments: 'good' patients, 'bad' patients and 'children', *Sociology of Health and Illness,* 5, 127–148, 1983.
20. Silverstone, R., Hirsch, E., and Morley, D., *Listening to a Long Conversation: An Ethnographic Approach to the Study of Information and Communication Technologies in the Home,* Brunel University, London, 1990.

21. Goffman, E., *Interaction Ritual*, Penguin, Harmondsworth, 1972.
22. Atkinson, P., Man's best hospital and the mug and muffin: An innocent ethnographer meets American medicine, in *Enter the Sociologist: Reflections on the practice of sociology*, McKeganey, N. and Cunningham-Burley, S., Eds., Avebury, Hampshire, 1987, 174–194.
23. Conaway, M., The pretense of the neutral researcher, *Self, Sex and Gender in Cross-Cultural Fieldwork*, Whitehead, T. and Conaway, M., Eds., University of Illinois Press, Chicago, 1986, 52–63.
24. Berridge, D. and Brodie, I., *Children's Homes Revisited*, Jessica Kingsley, London, 1988.
25. Davis, H., The social management of computing artifacts in nursing settings: An ethnographic account, Ph.D. thesis, University of Sheffield, England, 2000.
26. Pediani, R., A reflection on nursing time, *Journal of Advanced Nursing,* 28, 693–694, 1988.
27. Olesen, V. and Bone, B., Emotions in rationalizing organisations: conceptual notes from professional nursing in the U.S.A., in *Emotions in Social Life: Critical Themes and Contemporary Issues,* Bendelow, G. and Williams, S. (eds.), Routledge, London, 1988, p. 313–329.
28. Gold, R. L., Roles in sociological fieldwork, *Social Forces,* 36, 217–223, 1958.
29. Junker, B., *Fieldwork,* University of Chicago Press, Chicago, 1960.

chapter four

Narrative methods in qualitative research: potential for therapeutic transformation

Cynthia M. Stuhlmiller

Contents

0-8493-2075-5/01/$0.00+$.50
© 2001 by CRC Press LLC

Introduction

One of the best ways to obtain information is through an interview. While an interview can be rather straightforward, it can also be complicated, challenging, and rewarding to both the interviewer and interviewee. Whether the interview is aimed to find out about a potential job applicant or a person's problems or experiences, the interviewer attempts to build a story of the person through a conversational exchange that becomes a rich source for evaluation and understanding.[1]

Of the myriad types of interviews, the research interview and clinical interview share many common characteristics. Both employ face-to-face interview techniques that endeavor to capture firsthand accounts in order to uncover and understand human experiences and phenomena. While the purpose of the research interview is to gather information to answer a specific research question, and the clinical interview is used to diagnose or enable change of a person, it is possible for the reverse to happen. The clinical interview can generate information that can be used to theorize and the research interview can help a person clarify his or her situation and arrive at some helpful solutions or therapeutic outcomes. And, as with all interviews, a range of emotions will be elicited that can lead to both negative and positive outcomes for both the interviewer and interviewee.

In this chapter the focus is the links between the interview process used in narrative research methods and salutary outcomes for the interviewee. I begin with an overview of narratives followed by an examination of the interview process used in obtaining a narrative, including how and why this type of inquiry holds potential for therapeutic transformation or positive change for the interviewee. Drawing on examples from my research, I illustrate why the validation of telling one's story along with the interpersonal exchange between the storyteller (research participant) and listener (researcher) can create conditions that foster personal growth and wellness.

Nature of narratives

A narrative is a story of the events and individual experiences, told most often in a chronological fashion, for the purpose of understanding, conveying, and creating the meaning of experience.[25] People understand themselves through telling and hearing stories. As asserted by Gilbert,[11] "we live in stories, not statistics." Narration is the forward movement of a description of actions and events that makes the backward action of self-understanding possible.[9] Stories provide direct access to the richness of an encounter, including the situations, perceptions, and feelings that guided that person. Stories also serve to relate individual experience to the explanatory constructs of society and culture.[5] The study of multiple stories allows the discovery of connections that link people together and accounts for the differences between people.

People use narratives to explain the events that befall them. Stories create a sense of order out of chaos and give significance to an inexplicable event.[7] This can be seen in the case of Nazi Holocaust survivors who retell their stories of the horror they endured to succeeding generations to ensure that genocide will never again be tolerated.[10] In this and other narratives, bearing witness by providing testimony of terror, disgust, embarrassment, and defeat concretizes for subsequent generations a meaning for survival. Not only do narratives metaphorically create categories for interpreting events, but also they bind people and events into some intelligible pattern. People establish both order and social connections through the sharing of stories.[11]

Historically and across cultures, narratives have been used in societies to inspire and guide ethics, morality, and practice, and pass along tradition.[17] As asserted by MacIntyre,[16] only narrative descriptions of human life can render a system of ethics coherent and meaningful. Narratives teach powerful lessons about right and wrong, good and bad, and the right and wrong way to live. A story that offers moral engagement can exert a pull strong enough to direct a person toward good. Many examples can be found in parables, fairy tales, folk tales, and plays. Stories found in religious texts, such as the story of Job in the Old Testament, offers a description of an upright man whose faith in God enabled his survival in a test of repeated calamities. Everyday stories of human resilience and recovery can inspire the requisite hope and confidence needed for a person to act rather than give up.

Autobiographical narratives involve distance and intimacy. Distance occurs because the narrator is separated by the narrated events in time. This reflective stance could not be possible while the events were in progress. At the same time, the narrative is intimate, because it is a disclosure about the narrator and therefore there is a personal stake in how the story will be received.[9] This aspect of intimacy is important in narrative inquiry because the researcher, as an active participant in the interview process, is the main research tool (see Chapter 7). The researcher uses his or her interpersonal skills to set up the conditions for the narration to occur. Not only does the act of telling a story provide the study participant or narrator (to be referred to as such herein) an opportunity to order and make sense of his or her experience, the dynamics between the researcher and narrator has bearing on how the story is told and received.

The researcher and the narrative interview process

The researcher identifies who the best informants are for his or her study, makes contact, and negotiates a convenient and comfortable meeting time and place for the interview to occur. The initial contact and arrangements are important. Any tension between the researcher and narrator or discomfort created by an unsuitable environment will restrain the interview process.

The narrative research interview unfolds in three distinct phases that can be identified as the warm-up, the story, and the wrap-up. These phases follow logical temporal ordering of a narrative — beginning, middle, and end — as discussed by Carr.[8] The warm-up includes general introductions and chitchat leading to a review of the goals, purpose, and rules of the interview. During this phase, the researcher attempts to ensure comfort and reduce anxiety associated with the unfamiliar and unknown aspects of the situation. The researcher has an interest and responsibility to create optimal conditions that insure a productive and useful interview.

These first few moments of engagement are critical[21] because the researcher and narrator make initial judgments that determine the course of the interview. This phase helps the narrator solidify or reject the initial impression of the researcher, decide whether or not he or she likes the researcher, and how cooperative the narrator will be. The interpersonal transaction between the researcher and narrator is directly related to the quality and quantity of information revealed.

The story phase begins when the first interview question is asked. In narrative research, the questions are open-ended and typically general. "Tell me about yourself" or "tell me about your experience with such and such" is usual. How the researcher responds both verbally and nonverbally to the first answers further provides the narrator with a sense of the emotional climate to be maintained in the interview. This then enables the narrator to determine the level of personal depth and direction of self-disclosure that he or she is willing to take. The narrative includes the main story line and subplots related to the topic of the researcher's interest. The researcher may insert specific questions as probes to clarify and extend the narrator's story in order to gain as full an understanding as possible.

The wrap-up phase follows a natural course of storytelling. The narrator has reached the end of his or her story and the researcher determines that enough information has been obtained. At this point, the narrator assesses what he or she has revealed, how he or she feels about it, and considers how the information will be used. At the same time, the researcher is reviewing what has occurred, what information has been obtained, and what, if anything, is missing. During the wrap-up phase, it is the responsibility of the researcher to acknowledge and address any concerns that the narrator may have. Reiteration of confidentiality, use of data, follow-up, and any concluding remarks occur then.

Conditions for therapeutic transformation

The act of narration along with the interpersonal dynamics that occur during the narrative interview process can have a beneficial impact on the narrator. Telling a story about oneself can be an experience of healing and growth,[17] and participation in narrative research can be seen as a method of telling one's story.[11] It is the change (healing and growth) that can occur from telling a story and being heard by an interested listener that I define as therapeutic transformation in the narrative research context. I elaborate further on some of these elements in the following paragraphs.

Narrative research involves getting a story from an individual who is identified as having some knowledge or experience with the topic of the study. Thus the narrator is immediately identified as having expertise. This places the narrator in a position of relative power because he or she knows more about something than the researcher. In addition, the fact that a researcher finds the person's experience interesting, important, or relevant, suggests that the experience being investigated has some worth or merit. When someone asks, "tell me how you did that," or "what it was like," most people feel acknowledged as having something to offer or contribute. Even if the story is one of shame and remorse, the attempt to be understood rather than ignored, judged, or discounted often enables the person to feel valued.

Disclosure in order to gain information is the immediate purpose of a research interview. To get the story, the researcher must conduct himself or herself in a way conducive to fostering disclosure from the narrator. This may require the researcher to reveal something about himself or herself. Information-sharing and finding a common connection enables the narrators to realize that researchers, too, have experiences that they can relate to. This human-to-human connection can also lead to powerful moments of therapeutic change[6] when jointly shared emotions are explored in the research context.

The researcher facilitates disclosure by conveying empathy, respect, and genuineness. These characteristics help in uncovering rich data because the narrator feels that the researcher can be trusted with the story. Conversely, the researcher who is bored or gestures with raised eyebrows or facial expressions of disgust is signaling to the narrator his or her disinterest or disapproval of what is being revealed. In response, the narrator might respond aggressively, defensively, guardedly, or withdraw. Although there are no specific formulas for transmitting empathy, respect, and genuineness, when one individual feels acknowledged and understood by the other it is likely that acceptance, high regard, and care have been conveyed in a sincere, honest, and authentic way.

Probing and attentive listening by the researcher compel the narrator to provide specificity and detail to the account. When details are sought or discrepancies in the story emerge, a careful researcher will be interested enough to confront the narrator for clarification. When skillfully done, the confrontation is helpful to the narrator in sorting and ordering the story to make it be more coherent to themselves and to others.

The immediacy of the relationship enables the narrator to get spontaneous feedback that either validates or challenges the experience. If the researcher communicates warmth and a nonjudgmental attitude, the narrator will feel more comfortable in exploring his or her feelings and reactions in the presence of the researcher. This self-exploration enables the narrator to move his or her inner experience from the private sphere into the semi-public domain of validation and worth. The opportunity to work out puzzles and concerns can lead to therapeutic transformation during an interview.

This section outlined the factors that contribute to optimal disclosure in a narrative interview. The researcher sets up, directs, and facilitates the narrative disclosure through utilizing good communication skills that include: establishing a common bond, conveying empathy, respect, genuine acceptance, and regard, and listening, confronting, and clarifying. If the narrator experiences these elements, he or she will likely feel valued, important, accepted, listened to, and understood. These are the same conditions that have been identified as essential to creating a therapeutic relationship.[19]

Similar to a therapeutic relationship, the researcher listens to the person's narrative account and seeks to understand the process by which the person makes sense of their own behavior and the rules which govern their practices and social reality. Although the purpose of a therapeutic relationship differs from a narrative research interview, the interpersonal process is quite similar and therefore the outcomes, intentionally or unintentionally, may be similar as well. In the following section I draw on my own research experience to illustrate these points.

Narratives of disaster and therapeutic outcomes

My background

As Gilbert suggests,[11] those best suited to conducting narrative interviews are open to listening to the stories of others, with a willingness to consider ways of viewing the world that are different from their own. With a 20-year background of psychiatric/mental health nursing, I have a natural affinity toward a narrative research approach. Besides work experience in chronic and acute mental health settings, more than a decade of my clinical career was spent counseling Vietnam veterans and helping them come to grips with and reconcile themselves with the horror and trauma of war. During that time, I became interested in discovering the personal and cultural meanings and strategies that enabled some veterans to cope with their experiences in constructive self-enhancing ways. I felt the insight gained from this interest could be used to help others.

In effort to learn more about individual and collective stress and resilience, I engaged in a series of studies of recently occurring extreme stress events, specifically disasters.[22-25] I wanted to know how individuals respond to the experience of a disaster and its aftermath, how they interpret the event, their distress, and their coping.

What follows is a narrative description of that study. I have purposely adopted a narrative style for describing the research process and my experience as a researcher to illustrate the inextricable link between the researcher and the researched in narrative inquiry. The previous sections of this chapter outlined the foundational principles that I drew on to conduct this work. As you read my narrative, also consider what bearing my background as a clinician might have had on the study and its results.

The event

On October 17, 1989 an earthquake measuring 7.2 on the Richter scale struck the San Francisco Bay area. The largest loss of life occurred in Oakland California, where the collapse of a 0.76-mile stretch of double-decked roadway known as the Cypress Street viaduct resulted in the tragic deaths of 42 people. Both civilian and official rescue workers responded immediately. It is estimated that about 100 workers were on the scene at any one time.[22] Stressful working conditions included aftershocks, continual shifting of the compromised structure, tight working spaces, and pools of gasoline and fires. Within four days all survivors had been found and most of the bodies removed, but the process of dismantling the structure and retrieving all body parts continued for six weeks.

My disaster experience

It was a warm and sunny day in Northern California. World attention was focused on the San Francisco Bay area as preparations for game three of the World Series were getting underway at Candlestick Park. I was sitting in a classroom in Palo Alto, a city on the southern end of the San Francisco Bay peninsula. At 5:04 p.m., the ground began to rumble and the building began to shake. I watched through the large windows as parked cars bounced and street lamps swayed. Sensitized by several years' immersion in disasters as a Red Cross Disaster volunteer and as a doctoral student studying disasters, I knew immediately that this earthquake was a big one. I also knew that it would become the topic of the dissertation research that I had been planning to begin.

Researching people's responses to a disaster is an exciting, wearying, and ultimately rewarding experience — but an experience that in some ways directs the research in ways the researcher cannot. The most obvious quality of a disaster is that it gives no warning, and so the researcher must be ready to jump in with relatively little preparation and attempt to make sense of personal and social chaos. Much of my understanding of the event came later, after months of reflection, investigation, and follow-up discussions with people who became involved in the study. To give a taste of the event, I will describe my earthquake experience, impressions of my work at Cypress, the study, and why I believe that the narrative approach solidified therapeutic transformation for study participants.

The meeting I was attending continued on as if nothing had happened. I could hear the sound of sirens and the radio report of the collapse of the Bay Bridge. Following that announcement, the meeting quickly broke up and I made my way home. When I arrived home I found a mess but no major damage. That evening I checked in with other neighbors and helped them clean up. It wasn't until the next day that I heard of the Cypress Street viaduct collapse. The first phone call I received was from a colleague who was in Southern California and had heard of the collapse. He was putting together a psychological response team and wanted me to be involved. Despite my discomfort with capitalizing on human tragedy to pursue my research, this was the opportunity I had been waiting for. Now I could witness the rescue and recovery effort firsthand and revise my proposal according to the specifics of the operation.

Under the auspices of the Alameda County District Attorney, I joined a five-member mental health team assembled to provide psychological support to the rescue workers at the disaster scene for a period of five days. My role, determined by the team leader, was to provide critical incident stress defusing and debriefing to rescuers from what was considered by this world-renowned disaster expert to be a traumatic experience.

I found this an odd thing to do. As a Red Cross Disaster Action volunteer I had provided on-site psychological support to survivors and disaster workers. However, the idea of applying an across-the-board formalized intervention such as group defusing and debriefing seemed to overlook the fact that many people do well and actually thrive in extreme circumstances. Furthermore, it seemed inappropriate to force emotional reflection on work that was ongoing. I felt that coping with the immediate conditions of death and destruction might be aided by a lack of reflection. To me, the growing popularity of debriefing was a response to a manufactured rather than a real threat of widespread negative psychological consequences.

Since that time, the entire debriefing enterprise has been called into question. Several studies have suggested that debriefing is iatrogenic and can actually sensitize toward, rather than mitigate negative psychological outcomes.[4,12-15,18] Nonetheless, our charge then was to provide such help. One interesting aspect that has bearing on the research that I later conducted was the response of rescue workers to the presence of the team as well as the occupational titles I used when introducing myself.

As rescuers became aware that psychological help was at hand, they began to consider that the work they were involved in might be psychologically risky. When I introduced myself as part of the psychological team, people comically started twitching and exhibiting bizarre behavior in order to fulfill the expectations of needing help — "yes see, I am going nuts!" These reactions not only served to defuse anxiety and create a connection between the workers and psychological team, but also inadvertently instilled the notion that our services might be required. Rescuers then seemed compelled to dredge up stories of stress and negativity.

On occasion, over the next five days I introduced myself as a nurse with a background of working with Vietnam veterans. This title was met with interest and respect. The plight of Vietnam veterans had gained widespread recognition in the 1980s, largely through the diagnosis of post-traumatic stress disorder. Now experiences such as war, rape, the Holocaust, and disaster were understood as having a similar impact on the people involved. From this introduction, conversations began to include a more expanded view about the stress of rescue work. I was not a "shrink" looking for psychiatric symptoms of crazy people, but someone knowledgeable about what was considered to be normal stress reactions of normal people. Responses such as, "Oh, you're here to hang out in case someone needs help ... that's great!" were common.

Many of the rescuers were also veterans. They reminisced about wartime experiences and began to open up about what they were feeling about the rescue work. Because of my connection with veterans, I was easy to talk to and could understand. The devastation of the disaster scene also called forth the surreal and eerie images of a war zone. The disruption, discontinuity, and mixture of fear about what could have happened and what could still happen, and awed wonder at the earthquake's power, united not only veterans but all people at the scene to the extraordinary aspects of the encounter. The insider's perspective of being there and sharing in the awe became an invaluable aspect to the study that I conducted later on.

Also, as is typical for nurses in clinical, social, and research settings, people felt free to discuss their physical health status. "Do you want to see my scar?" "Did I tell you about my bypass surgery?" "What do you do for constipation?" were typical of the conversations that nurses often find themselves in. The belief is that nurses have seen and done it all so they can handle anything you tell them. This was true at Cypress. I seemed to get more detailed stories of the gruesome conditions from rescuers than did my other colleagues. This could be attributed to my personality traits; however, I noticed that the way in which I presented myself brought about different reactions that corresponded to my title and experience. This was underscored subsequently in my study.

To conclude or wrap-up my narrative, the five-day encounter with the Cypress operation was profound. I was privileged to witness the dramatic acts of men and women who transcended their known abilities to save lives and retrieve bodies from the collapsed structure. This has had a lasting effect. To this day, I continue to be struck by the human capacity to jump into dangerous situations and help others. The ability to go on and learn from tragedy is a course of continual wonderment and hope.

Study participants

Six months following the event, I obtained human subject approval from my university to study the rescuers. Of all the rescuers responding to the Cypress collapse, the following four groups were selected for the study based on the fact that they were highly exposed to death and played a major role in body rescue and recovery. They were firefighters (n = 15), military pararescuers (n = 6), coroner investigators (n = 6), and transportation workers (n = 15). The number of informants was determined by the numerical ratio of responders from each agency as well as data saturation — when the stories became repetitive and no new information was gleaned.

Data collection techniques

Three major data collection techniques were used in the study: document analysis, participant observation, and taped narrative interviews. Document analysis involved reviewing all available related information about the earthquake event and the individuals who were the focus of the inquiry. Newspaper, television, video, periodicals, books, after-incident reports, and agency records were used. Also, from the moment of disaster I kept a notebook of my personal reactions, a journal of my work at the disaster site, as well as related letters, poems, photos, and journals that people offered to me. All these data sources supplied the background for understanding how members of a community and structures within a community create conditions for the experience and expression of personal accounts.

Given the disillusionment with the involvement I experienced at the disaster site and driven from the one-sided concern for symptom formation, I only considered research methods that might enable a more balanced picture to emerge. I also had no interest in taking a removed approach such as that of a paper-pencil survey, given the personal connection I had to the event. I therefore decided on narrative interviews as the most appropriate method.

Agency supervisors released the names of individuals who were most involved in the rescue and recovery effort. Informants were contacted by phone or in person and a meeting time and place of their convenience was arranged. Prior to each interview, an overview of the study was explained, a demographic data sheet filled out, and an explanation given of what to do in case of negative aftereffects from the interview. Withdrawal from the study at any point was assured, as well as a copy of the study findings.

The narrative interview followed the temporal format described previously — the warm-up, the story, and the wrap-up. The main question was, "Tell me about yourself, your experience of disaster and rescue work." The question was specifically designed to be as broad and general as possible to allow each person to talk freely about anything that had meaning or relevance to their experience. As people told their stories, they described their occupational background, personal experience of disaster, how they became involved in the rescue operation, what they thought, how they felt, what they did during and following the disaster, what helped, and what they learned. This design also enabled rescuers to describe both the common and unique experiences as they related to the context of the work that occurred at the collapse site.

Follow-up interviews were conducted at 18 months post-event to gain further information about the lasting memories, consequences, and importance of the rescuers' experiences over time. These interviews provided consensus validation of the research findings and extended the established connection I had made with each person. Rescuers were most appreciative and impressed that a researcher would go to such lengths to provide research findings directly and to maintain interest in them as individuals.

Narrative analysis

Narrative analysis was guided by the interpretive phenomenological strategies developed by Benner.[2,3] The analytic procedure follows a four-stage process of identifying themes, exemplars, and paradigm cases that are used to form the descriptions of individual experience. The first stage involved reading the narratives as a whole in order to describe aspects of the person's experience. Fieldnotes and impressions from the interview were additionally used to form beginning interpretations. By this method, the relative importance of the event to the individual's life was assessed and summarized.

In the second stage, thematic analysis, all interview excerpts relating to the major themes of the earthquake experience were extracted verbatim from the interview transcripts. An in-depth study was made in order to identify these themes and their variations.

The third level of analysis involved locating exemplars of experience that appeared to be especially meaningful. Exemplars, or specific examples of patterns that depict the appraisal and coping process, were culled from each interview and presented as vignettes of experience.

The fourth stage, identification of paradigm cases, consisted of reading and rereading the case material generated in the previous stages for similarities and differences to see if cases could be grouped by similar meanings. This phase looked for themes evident in all cases that were presented through interview narratives. Paradigm cases illustrate particular patterns and variations of how the disaster and rescue work were experienced.

Findings

A more detailed report of the findings can be found in References 23 to 25. Overall, I found that the narrative approach enabled participants to discuss the meanings, issues, situations, and concerns that were most critical to them. Meeting with firefighters, transportation workers, coroners, and military pararescuers (PJs) taught me not only what rescue work at Cypress entailed, but also about the impact that work practices and meanings have on the lives of these persons. The people I interviewed had strong ties to their work — ties formed and strengthened by indoctrination, other milder forms of socialization, and practices specific to their occupations.

The rescue-related events they defined as stressful involved threats and challenges to the expectations, mores, and goals of each occupational practice in relation to the disaster work. The value and meaning of the work and personal understanding of role obligations also helped to determine the form of involvement at Cypress and the definition of stress.

Potential for therapeutic transformation

This concluding section links the disaster study with the general therapeutic value of narratives, pointing to the specific circumstances of the disaster event that provided the context in which transformation occurred. I have organized the discussion into topics of: participant observation, significance of the researcher's position, the power of asking for a story, narrative vs. other approaches to research, and the transformative power of narratives.

Participant observation

Although I had not previously met most of the rescuers I interviewed for the study, the fact that they knew that I had been at the disaster scene created an instantaneous connection between myself and each narrator. The warm-up phase contained so many familiar exchanges that it was like a homecoming. Connected by a "flashbulb memory" (memories of the circumstances in which people first learn of a very surprising or emotionally arousing event such as the shooting of President John F. Kennedy), the earthquake event, and rescue operation provided a powerful reference point that set the narrative interview in motion. Each and every narrator made comments such as, "Remember when this happened?" or, "I don't need to explain this to you." This common bond contributed to narrative disclosure, which is also known to be therapeutic. Respect, empathy, and genuineness were experienced as mutual because of the shared participation.

Researcher's position

As discussed earlier, the title and role held by a person carries an expectation that influences relationships. At Cypress, my role as nurse, Vietnam veterans counselor, and psychiatric team member generated different responses. As a researcher, I was in yet another position. I was now interested in finding how people interpreted their encounter with disaster and how they coped. I was concerned with getting an overall narrative of experience and not just symptoms of physical or emotional reaction. This allowed narrators to determine and reveal the aspects of experiences that they thought were most salient rather than follow some limited pattern of questions that I thought was important.

This stands in contrast to other research techniques that strictly guide what information is elicited and therefore what information has value. From the open stance of narrative, I obtained a much fuller and richer story of the encounter than I would have from a structured interview. Spoken accounts allow the speaker to give more details and include concerns and considerations that shape the person's experience and perception of the event.[3] Therefore, narrative accounts are meaningful accounts that point to what is perceived, what is worth noticing, and what concerned the storyteller.[20]

As a researcher collecting stories, I was not necessarily in a hierarchical position to the narrator. Although some researchers would argue that there is no way to avoid the hierarchical character of the research encounter, I was not making a clinical assessment or offering any expert evaluation. Perhaps the fact that I, too, had been involved in the disaster offered a more equitable platform from which the investigation took place. Also, what was revealed to me was anonymous and could be withdrawn without repercussions.

While most narrators want to please the researcher or portray themselves in a positive light, the rescuers took this reflective opportunity to check out their concerns. For example, one firefighter was explaining his guilt about receiving an award for heroic behavior. As he attempted to work out his guilt in the narrative, I reminded him of an experience he had told me earlier in the story. This information, brought forward, seemed to clear up his questions and concerns. Through careful listening, the researcher can help narrators connect pieces of a story into a coherent whole.

Many rescuers wanted to know how their story compared to others. It was easy to point out similarities and differences without violating research ethics. As I described in the previous section, members of each occupational group had a particular, shared way of appraising and coping with situations at Cypress. I was therefore able to point out the commonalties and differences in a general way. This was therapeutic because it reassured the narrators that they had much in common with their peer groups and yet their group was distinctive from others. A solidified sense of self and membership in an occupation emerged as the narrators described their rescue experiences from their job perspective.

The power of asking for a story

When approached, many rescuers remarked, "Why ask me, I didn't do so much," or, "Other people have better stories than me." When I responded that they indeed had a story to tell and that I wanted to hear it, a transformation occurred. Rescuers who felt uncertain about the value of their contribution at Cypress contemplated for a moment and then replied, "Oh well, maybe I did do something important. If you think I have something to say, perhaps I will add something useful to your study." At the completion of their narrative, several of the rescuers who had been hesitant to offer their story said, "Gee, I am actually proud of what I did out there. When you think about it, I actually helped undo a tragedy." The simple acknowledgment that each person had an important story was empowering. It contributed to a sense of personal pride in involvement. Furthermore, the person's narrative would now become part of a written document — a testimony of what occurred at a particular moment of history.

Rescuers all wanted copies of the book that resulted from the study. When that eventually occurred, I watched people combing through the text looking for their words. "Hey, that's me, wow, I am in print." This seemed to concretize their experience in a most positive way.

An important aspect regarding the narrative approach taken was to conclude each interview with wrap-up questions such as, "What did you learn about yourself from your encounter?" and, "What advice would you have for others?" This directed narrators to consider the positive or growth-promoting aspects of the experience and enabled such thoughts to linger after the interview.

Narrative vs. other approaches

The therapeutic impact of the narrative approach used in this study can be judged against other studies that were conducted about the Cypress rescue and recovery operation. One researcher provided rescuers with a battery of questions on paper and asked them to rate their level of distress regarding certain aspects of the disaster event. The questionnaires were dropped off at various agencies without any face-to-face involvement with study partic-ipants. Several people who were also part of my study were upset by the questions which included, "How upset were you seeing maggots on decom-posed bodies?" and "How upset were you seeing rats eating disfigured and dismembered bodies?" In fact, because only a few people had seen the maggots or rats, this line of inquiry led to anxiety and uncertainty of several rescuers who now began to rethink what they had seen.

One rescuer said to me, "I didn't know all of this went on. I rescued a few live people and was stoked about it, now feel bad about other rescuers who must have horrific nightmares ... now I am beginning to have them too." Another rescuer said, "Now I wonder if what I saw moving around on a body, which I thought was the wind rustling paper, was actually a rat." Without any personal contact or further explanation by the researcher, res-cuers began to doubt their role in the research and conjure up concerns about the negative impact of their disaster encounter. One rescuer remarked, "I filled out a questionnaire and turned it in but I wonder what has happened to it. I don't know what my answers mean nor do I know if I am OK or not."

From these examples, narrative research approaches can be judged as having more potential for health promotion. The researcher who uses nar-rative methods can provide on-the-spot feedback, validation, and reassur-ance, as well as suggest further help for the troubling emotional aftermath from the event or research process if deemed necessary. Of course, much of the positive potential of the narrative interview is related to the skill, sensi-tivity, and overall awareness of the researcher. Good clinicians tend to make good narrative researchers.

However, as is pointed out in Chapter 7 of this volume, unpleasant, upsetting, and traumatic memories can also be triggered by qualitative research questions which can lead to negative and damaging outcomes for the interviewee. It is therefore essential that researchers using narrative inquiry undergo practice and training in critical reflection and have some level of confidence to deal with a range of emotions and responses that occur.[3] For example, taped recordings of practice interviews will enable the researcher to evaluate his or her anxiety, silence, avoidance, and inability to pick up obvious and subtle cues. Critique by other researchers will help to identify blind spots.

Research methods that seek to discover and rate only negative encounters can lead to the same iatrogenic effects that have been associated with psychological debriefing. A one-sided focus on eliciting narratives of negative outcomes can serve to reinforce vulnerability and may shortcut the individual's natural restorative capabilities. A deficit approach taken in a narrative interview can overshadow positive outcomes and may suggest to the individual that a certain amount of anxiety should be expected or that there are some standard responses to which they should resort.

One rescuer told me that he felt guilty because he didn't feel sad or upset about his work at Cypress. Feeling bad was the "normal response" that the psychological debriefing team at Cypress told him he could expect. He even showed me the symptom checklist he had been given that outlined the expected anxiety he might experience. Instead, he felt good about his participation, and his confidence to work in disasters had increased. Because he felt good, he considered it meant that he wasn't normal.

Transformative power of narratives

The rescuers described their experience at Cypress within the context of their life history. As they did so, they explained the connectedness of their thoughts, feelings, and actions in relations to past experiences and their current situation. Their stories uncover reasons for participation: what they did, how they felt about it, how they interacted with others, and how their experience has affected their lives.

Several reasons for preserving rescuer narratives emerged from this project. The informants' narratives reflect courage and human possibility as well as fear and a sense of vulnerability. Their accounts provide visions of the value and integrity of life. The most striking thread throughout the story was the reservoir of shared cultural meanings related to personal commitment, dedication to saving human lives, avoiding suffering, facing death, and comforting families, from which these people drew to help their fellow beings.

In a pluralistic society that struggles in search of common meanings and mores, these stories capture a sense of commonality and solidarity that prevailed in one particular moment in history. As Taylor[26] suggests, the search for mortality extends beyond people's obligations to other people, to include such confronting questions as: how am I going to live my life, what kind of life is worth living, and what constitutes a rich and meaningful life? In the telling of their stories, informants answered these questions as they made sense of their background experiences and self-understanding in relation to the rescue work.

Perhaps this is why the study of Cypress yielded such stories of personal and collective enrichment. There was no further death or destruction during the rescue operation and the event itself was unique and unifying. The fundamental foundations about life's certainty had been shaken. Each one of us gained new insight and were forever changed or transformed by our disaster experience. When new insight or positive meaning can be found in situations of death, destruction, distress, and suffering, therapeutic change occurs.

For some, the recognition of one's mortality from the fact that, "you can be wiped out in a moment," ties in with new insight, as does the sense of having a better perspective on life, knowing what is important and not "worrying about the little things so much." For others, the experience had given them a clearer sense of themselves. One rescuer learned that, "I am not made of steel emotionally." Another took the opportunity to "sit down and take an inventory of me."

This clearer understanding extended as well to the rescuer's occupations. On a basic level, the experience was kind of a proving ground that validated the individual's training and ability. On a deeper level the Cypress collapse confirmed the importance of their work. "It makes you realize that the sacrifice involved with this job is worth it. It substantiated that what I do has a purpose."

Several rescuers noted a new or renewed sense of a larger force: some finding new respect for Mother Nature, and others finding that their religious faith had been strengthened. This sense was often accompanied by a recognition of their relative smallness in the face of this larger force and thus of their relative lack of control over the events in their lives.

The event, not necessarily the narrative research process, initiated new understanding and personal transformation. However, each and every research participant remarked about how useful telling their story was. They felt clearer about their participation and even closer and more involved with a tangible sense of purpose through committing their story and contributing to the research. Undoubtedly, the narrative approach deepened and legitimized experiences by linking these insights to personal values, beliefs, and one's purpose for living.

Conclusion

Narrative research methods can exert a therapeutic effect because they have the strength, power, and generative capacity to uncover and foster growth, possibility, and relatedness. It has been pointed out in this chapter that telling a story of the experience, combined with the interpersonal exchange between the researcher and narrator, connects people to their experiences, their worlds, and each other. It is indeed a privilege to listen and bear witness to the encounters of fellow humans. With this privilege comes responsibility — not only to provide faithful accounts of the testimonies told but also to guard against the risks and vicissitudes of telling and listening. The title of this book affirms that the emotional nature of qualitative research supplies the researcher and the researched with untold challenges and rewards. It is the potential for overwhelming reward that keeps researchers like myself engaged in such endeavors.

References

1. Barker, P., *Assessment in Psychiatric and Mental Health Nursing: In Search of the Whole Person*, Stanley Thornes, Cheltenham, U.K., 1997.
2. Benner, P., *Stress and Satisfaction on the Job: Work Meanings and Coping of Mid-Career Men*, Praeger, New York, 1984.
3. Benner, P., *Interpretive Phenomenology: Embodiment, Caring, and Ethics in Health Care and Illness*, Sage, Thousand Oaks, CA, 1994.
4. Bisson, I., Jenkins, P., Alexander, J., and Bannister, C., A randomized clinical control trial of psychological debriefing for victims of acute harm, *Br. J. Psychiatry*, 171, 78–81, 1997.
5. Brody, H., *Stories of Sickness*, Yale University Press, New Haven, CT, 1987.
6. Brown, V., Psychotherapists' Strong Reactions: An Empirical, Phenomenological Investigation, Doctoral Dissertation, Dusquesne University, Pittsburgh, 1986.
7. Bruner, J., *Acts of Meaning*, Harvard University Press, Cambridge, MA, 1990.
8. Carr, D., *Time, Narrative, and History*, Indiana University Press, Indianapolis, 1986.
9. Churchill, L. R. and Churchill, S. W., Storytelling in medical arenas: The art of self determination, *Lit. Med.*, 1, 73–79, 1982.
10. Felman, S. and Laub, D., *Testimony: Crisis of Witnessing in Literature, Psychoanalysis, and History*, Routledge, London, 1992.
11. Gilbert, K., Taking a narrative approach to grief research: Finding meaning in stories. Unpublished paper presented at the 5th Int. Conf. Assoc. Death Education and Counseling, Washington, D.C., 1997.
12. Griffiths, J. and Watts, R., *The Kempsey and Grafton Bus Crashes: The Aftermath*, International Design Solutions, East Lismore, Australia, 1992.
13. Gist, R., Lubin, B., and Redburn, B., Psychological, ecological, and community perspectives on disaster response, *J. Personal Interpersonal Loss*, 3, 25–51, 1998.
14. Hobbs, M., Mayou, R., Harrison, B., and Worlock, P., A randomized clinical control trial of psychological debriefing for victims of road accidents, *Br. Med. J.*, 313, 1438–1439, 1996.
15. Kennardy, J. A., Webster, R. A., Lwein, T., Carr, V., Hazell, P., and Carter, G., Stress debriefing and patterns of recovery following a natural disaster, *J. Traumatic Stress*, 9, 37–49, 1996.
16. MacIntyre, A., *After Virture*, University of Notre Dame Press, Notre Dame, IN, 1984.
17. McAdams, D., *The Stories We Live By*, William Morrow and Co., New York, 1993.
18. McFarlane, A., The longitudinal course of posttraumatic morbidity: The range of outcomes and their predictors, *J. Nerv. Ment. Dis.*, 176, 30–39, 1988.
19. Rogers, C., *On Becoming a Person*, Houghton Mifflin, Boston, 1961.
20. Rubin, J., Impediments to the development of clinical knowledge and ethical judgment, in *Expertise in Nursing Practice: Clinical Knowing, Clinical Judgment, and Skillful Ethical Comportment*, Benner, P., Tanner, C., and Chesla, C., Eds., Springer, New York, 1995.
21. Shea, S. C., *Psychiatric Interviewing: The Art of Understanding*, 2nd ed., W. B. Saunders Company, Philadelphia, 1989.
22. Stuhlmiller, C., An Interpretive Study of Appraisal and Coping of Rescue Workers in an Earthquake Disaster: The Cypress Collapse, Doctoral dissertation, University of California, San Francisco, 1992.

23. Stuhlmiller, C., Rescuers of Cypress: Work meanings and practices that guided appraisal and coping, *West. J. Nursing Res.*, 16(3) 268–287, 1994.
24. Stuhlmiller, C., Narrative phenomenology in disaster studies: Rescuers of Cypress, in *Interpretive Phenomenology: Embodiment, Caring, and Ethics in Health and Illness*, Benner, P., Ed., Sage, Thousand Oaks, CA, 1994, 323–349.
25. Stuhlmiller, C., *Rescuers of Cypress: Learning from Disaster*, Peter Lang, New York, 1996.
26. Taylor, C., *Sources of the Self: The Making of Modern Identity*, Harvard University Press, Cambridge, MA, 1989.

chapter five

Mirrors: seeing each other and ourselves through fieldwork

Lisa M. Tillmann-Healy and Christine E. Kiesinger

Contents

0-8493-2075-5/01/$0.00+$.50
© 2001 by CRC Press LLC

> Every insight was both a doorway and a mirror — a
> way to see into their experience and a way to look back
> at mine.
>
> — Michael Schwalbe, *The Mirrors in Men's Faces*[32]

Introduction

In this chapter, we discuss and show the dialectical relationships between
fieldwork, identity, and emotional experience. In the course of our project,
each of us (1) has written autobiographically about her struggles with
bulimia, (2) has interviewed the other about her experience, (3) has written
biographically about the other's experience, and (4) has read and responded
to the other's texts. By engaging in these varied activities, we have come to
think of fieldwork and field relationships as mirrors, as ways of seeing both
others and ourselves.

Our project combines autoethnography[18-19] and narrative ethnography[35] —
two unconventional ways of practicing fieldwork and of writing about fieldwork experience and relationships. All forms of ethnography aim to increase
cultural understanding. In traditional ethnography, researchers engage in
participant observation, usually in cultures to which they don't belong.
Autoethnographers, in contrast, are participant observers of their own experience. In narrative ethnography, researchers move from participant observation to the *observation of participation*.[35] Here, cultural analysis emerges
from the character and process of the ethnographic *dialogue* between
researcher(s) and participant(s). Both auto- and narrative ethnography
require us to think in nontraditional ways about the role of the researcher,
the look and feel of ethnographic writing, and the position of readers.

Traditional ethnographers are trained to approach research "objectively."
Their emotions often are seen as "contaminating" the research process; therefore, these must be controlled as much as possible. In auto- and narrative
ethnography, however, emotions are seen not as biases to be eliminated but
as unique sources of insight to be valued and examined, and to be featured
in ethnographic texts. Because of this, such fieldwork demands a higher
degree of reflexivity than does traditional ethnography.

Auto- and narrative ethnographers also are more self-consciously political than more conventional researchers. We reject neutrality as a synonym for estrangement that is both unachievable and undesirable.[24] Instead, we take a *purposefully ethical stance* toward research and participants.[27]

As writers, moreover, auto- and narrative ethnographers strive to compose texts quite unlike standard academic works. Our ethnographies draw from fields as diverse as communication, anthropology, sociology, psychology, medicine, literary criticism, cultural studies, media studies, the visual arts, popular culture, and literature. Such radical intertextuality renders auto- and narrative ethnography interdisciplinary, transdisciplinary, and even counterdisciplinary.[9]

Our writing also has an unorthodox "feel." Auto- and narrative ethnographers privilege embodied experience as both a subject of research and even as a method of inquiry.[6] Our texts move beyond the ethnographic gaze, emphasizing what is smelled, heard, and felt as much as what is seen.[24] If successful, the work gives readers, in Stoller's[34] words, "a *sense* of what it is like to live in other worlds, a taste of ethnographic things."

To produce an evocative auto- or narrative ethnography, we employ techniques more often associated with fiction and new journalism than with social science. These include scene setting, thick description, (re)constructed dialogue, foreshadowing, dramatic tension, and temporal shifts.[8] Inviting readers inside fieldwork experience and relationships, our texts take shape in one or more forms, such as ethnographic short stories,[15] ethnographic fiction,[1] poetry,[2,28] ethnographic drama,[16] and layered accounts.[30]

The criteria by which we judge these ethnographies are different from those used to evaluate orthodox social science. Moving from factual truth to narrative truth,[4,33] such projects can be assessed by their personal, relational, and cultural *consequences*.[24] Says Robert Coles,[7] "there are many interpretations to a good story, and it isn't a question of which one is right or wrong but of what you do with what you've read." The best stories, according to Bochner,[4] expand our sense of community, deepen our ability to empathize, and enlarge "our capacity to cope with complicated contingencies of lived interpersonal experience."

Our readers, finally, occupy an unconventional position. Auto- and narrative ethnographers write for those who want to be engaged on multiple levels — intellectually, emotionally, ethically, and aesthetically; to confront texts from their own experience; and to participate as co-producers of meaning in ongoing conversations. Such ethnographies embrace, in Denzin's terms,[8] a "dialogical ethics of reading."

Because auto- and narrative ethnographies remain open-ended, encouraging multiple interpretations, readers are invited to offer critical responses in the form of negotiated and subversive readings. In the process, texts become sites of political negotiation. Ideally, by interacting with the work, readers find something to take in and use, both for themselves[7] and for social change.[8]

"Mirrors": a confessional tale

In the pages that follow, we provide a detailed description of our ethnographic project in the form of a multistep, co-narrated "confessional tale." According to Van Maanen,[38] this is a highly personalized account that seeks to demystify the research process and to portray the impact of fieldwork on the fieldworker. In a confessional tale, one writes not as a distanced, disembodied, and "objective" scientist, but as a human ethnographer who attempts to make sense of (and cope with) the research experience. Unlike a "realist tale,"[38] whose authority is based on the strength of its general claims, a confessional tale gains credibility through the evocative power of its particularities. As you engage the details of our project, we hope that you will sense and *feel* how qualitative research on emotional experience can transform the lives and identities of researchers, respondents, writers, and readers.

Step 1: Christine writes about herself

I first wrote about my history as a bulimic woman in 1992. At that time, my sister Julie and I were realizing that both of us had problematic relationships with our bodies and with food.

I was taking Emotional Sociology, a graduate course in which I was introduced to autoethnography. Julie and I decided to use my class project as a means to understand our own and each other's bulimia and to raise others' consciousness about eating disorders.

Using systematic introspection,[11] we immersed ourselves in memories related to body image and eating. Each jotted down descriptions and reflections, and these became the fieldnotes from which she composed ethnographic short stories[15] and ethnographic poems.[28] In the final version, "Writing it Down: Sisters, Food, Eating and Our Bodies,"[26] I arranged our stories and offered theoretical commentary on the narrative turn in the social sciences.[4] This turn posits narrative as a communication paradigm,[22] as a method of inquiry,[17,31] and as a mode of representation.[12,14,17]

In "Writing it Down," Julie and I constructed our eating disorder as a response to our fear of becoming obese (like many of our relatives). Looking back, this paper merely scratched the surface of what later would be uncovered as the deeper, more complex meanings of bulimia in my life.

Nevertheless, authoring this paper was a powerful and intense experience. Thoughts and feelings emerged almost effortlessly from my mind and body. For the first time, I was able to voice and release some of the rage, grief, and ambivalence surrounding my eating disorder.

After I wrote through an episode, I would examine the words on the page:

> *Ex-Lax®. They are easy to chew. They feel soft and*
>   *soothing in my mouth.*
> *Taste like chocolate.*

After chewing a few, there is nothing to do, except wait.
Wait until my stomach cramps and then, ruptures,
  explodes.

And then it all spills out of me,
  empties me and then
ruptures,
  explodes,
again and again and again.

Startled by my own graphic details and stark tone, I often was left thinking, "Is this really me?"

Step 2: Lisa reads Christine and Julie's autoethnography

In the spring of 1994, I enrolled in Art Bochner's graduate seminar on narrative inquiry. Late that term, Art assigned the paper Christine co-authored with her sister.

I opened "Writing it Down" tentatively, unsure that I wanted to confront these women's stories, stories that might force me to confront *my own* secret life. Unbeknown to my family and friends, I had been binging and purging since age 15.

I began to read, and the images took hold of me:

Pre-made Pillsbury® cookie dough, chocolate chip.
I eat spoonfuls of it, several spoonfuls.
It is rich, sickenly sweet, filling.
It makes me feel full inside.

My eyes fixed on that last sentence. "It makes me feel full inside." Yes, I thought, that's what food does for me when nothing else will: *it makes me feel full inside.*

"Writing it Down" drew me into Christine and Julie's private wars. In both women, I saw reflections of my long-hidden self. After I finished reading, I rushed to my computer and pounded out seven pages of reactions and memories triggered by their paper.

Step 3: Lisa writes about herself

In my weekly journal entry for Art Bochner's course, I followed Christine and Julie's style by creating autoethnographic snapshots of my bulimic experiences. The episodes I reconstructed showed my early weight consciousness, my first purge, and my contradictory desires to reveal and conceal bulimia.

Writing about my conflicted relationships with my body and with food felt much like a purge — painful yet relieving. Many tears fell onto my keyboard in those hours, but I remained driven, at last, to unleash some of the pain and guilt. I photocopied the entry for Art, who served as both my advisor and Christine's. After examining it, he asked my permission to share what I wrote with Christine, who was working on her dissertation, a collection of evocative life histories based on interviews with anorexic and bulimic women. A bit anxiously, I agreed.

Step 4: Christine reads Lisa's journal entry

When Art gave me a copy of Lisa's entry, he told me that I should consider inviting her to participate in my project on women with eating disorders. Reading her words, I knew why.

Her descriptions were vivid and raw. I felt privileged to witness such incredibly personal scenes. At the same time, I felt like a voyeur. Until then, I only had known Lisa as a gifted and respected colleague. Was I ready to confront this messier picture that so clearly exposed her anguish?

As a writer, I understood that Lisa's entry portrayed some of her deepest fears and longings; but as a reader, her stories also opened a door to *my own* fears and longings. This struck me most powerfully when I read Lisa's recollections of the night she told her lover about her bulimia:[37]

> My cheek presses against his chest. His breathing shifts
>   over from consciousness to sleep.
> *Do it, Lisa. Don't wait.*
> "Douglas?" my strained voice calls out.
> "Umhmm."
> "There's something I need to tell you."
> *Probably not a good opening line.*
> "What's that?" he asks.
> "I know I should have told you this before, and I hope
>   you won't be upset that I didn't." Deep breath.
>   Swallow.
> *It's okay. You're doing fine.*
> "What is it?" he asks, more insistent this time.
> …"Oh, god, Douglas…I have been bulimic … since I
>   was 15."
> *It's out there. You said it.*
> He pulls me closer. "Well, how bad is it now?"
> "It has been much worse."
> "That's not what I asked."
> "It's not that bad."
> *Liar.*
> …"I'm really glad you told me," he says as I start to
>   cry. "I love you, Lisa. Tell me how I can help you.
>   Please."
> *You just did. You can't imagine how much.*
> He pulls me close, stroking my hair until I go to sleep.

I scanned that episode over and over. The scene reminded me of those times when, after making love, I laid still against my lover, watching the steady rise and fall of his chest. How desperately I wanted to tell him, to whisper to him about my unspeakable rituals — the binges while he was away, the purges while he slept. Those were unbearable stretches of time, and just when I was on the verge of saying something — anything — I'd hear the deepening of his breath and know that he was asleep. I'd turn over and away from him, alone again in the silence.

I envied the courage it must have required for Lisa to speak out in that quiet moment in the dark. I envied her lover's warm, gentle response. I hungered for the same kind of response and understanding from those in my life.

Step 5: Christine interviews Lisa

Christine's story*

I took Art's advice and asked Lisa to participate in my study. She agreed, but months passed before I scheduled an interview. The delay was intended.

I always had looked forward to meeting with my participants. Our sessions involved "interactive interviewing,"[20] an interpretive practice that privileges the lived, emotional experience of both respondent and researcher, that requires such intense collaboration that these roles blur, and that demands deep and prolonged interpersonal contact.

But as it came time for a session with Lisa, I was *scared*, scared of what she would reveal, scared of what *I* might have to reveal, scared of the light her story could shed on mine. Of all my participants, Lisa was most fully my peer. She reflected so many of my selves: professional and personal, public and private. Perhaps that's why, for the first time since beginning my study, I felt the need to conceal certain parts of my story.

When the afternoon of my first interview with Lisa finally arrived, I was incredibly nervous. My heart jumped at the sound of the doorbell.

I opened the door. Smiling, she appeared fresh and radiant. Motioning for her to come inside, I offered a beverage and asked Lisa to sit where she would feel most comfortable. As I squeezed lemon into our iced teas, I noticed that my palms were sweating.

We chatted at length about graduate school, flitting from one issue to another. I began to wonder if we would continue the small talk forever, avoiding the "real issue." At last, when there was a brief break in the conversation, I forced myself to look her in the eyes and ask if we could get started.

"Let's do it," she replied. With a trembling hand, I reached over and pressed "record."

In the minutes that followed, Lisa and I proved masters at intellectualizing bulimia. We abstracted it, dissected it, and theorized it, rendering our

* Portions of this section are adapted from Reference 20.

experiences almost totally devoid of emotion — a remarkable contrast to how each of us had written about her eating disorder.

I struggled through our session. I asked some questions, shared some experiences, and probed her for details. But it fell flat. We connected in a lot of ways: small-town upbringings, troubled adolescences, obsessive relationships with men, attractive yet self-conscious mothers. But there was so much that wasn't being said, so many doors we were leaving closed.

Though she never let me in, I knew what went on behind the closed doors in Lisa's life. I knew because I knew what went on behind *my own*. How, I wondered, could we open our doors?

I tried one last key. "Tell me about your father," I said, hoping their relationship might shed some light on Lisa's secret life. In my fieldwork, I had listened to bulimic women talk of physically, emotionally, and even sexually abusive fathers, distant or absent fathers, and critical and overprotective fathers. Through their accounts, I had begun to see my father's rages (many of them occurring at the dinner table) as an important plot line in the evolving story of my disordered relationship with food.

To my surprise, Lisa *smiled* at my request.

I tried to be happy for her as she laughed about dancing at father/daughter banquets and being "daddy's special girl." But instead, I descended into childhood grief. How I longed for a "daddy" like the one Lisa described — a daddy with strong, warm hands and a soft, gentle smile; a daddy who dried tears, made his daughter laugh, and kept her secure. This point of disconnection between Lisa and me was an important one to explore. If Lisa's father wasn't part of her bulimia's plot, I needed to understand who and/or what was. But I was too exhausted, too wounded, to push further.

I ended our session. When Lisa left, she looked so together, so untouched. Compared to hers, my life seemed a total mess. Then again, if her life was so ordered, what led Lisa to this "disorder?" She talked about wanting to be thin and being self-destructive, but did that account for the kind of control bulimia sometimes had over her life?

Months later, I interviewed Lisa a second time in a small German restaurant. My goal was to push for elaboration. I told her of my inability to access the emotionality of her story, and for the first time, Lisa began to open up. Reflecting on this session, I wrote the following:*

> *"Lisa," I recall, "I once asked how you felt — how you feel — when you're about to purge. You explained that you felt like a little girl. Do you remember telling me that?"*
>
> *"No, no," she answers softly, almost shyly. She pauses, looking at me quizzically, and asks, "What do you make of me saying that?"*
>
> *"I don't know."*
>
> *She says nothing.*

---

* Adapted from Reference 25.

With hesitance, I push on. "Of all the accounts of women I've interviewed, yours has been the most difficult for me to understand and to write. You aren't debilitated by bulimia. You're a successful, functioning woman. I suspect that all stories of this disorder are tragic somehow, but I find myself unable to behold your tragedy. With my other participants, I saw and felt their pain, but I don't see or feel yours. Can you tell me about — can you show me — the private side of your bulimia?"

It has been excruciating for me to reveal these things to Lisa. Why does pressing her to uncover the contradictions, the confusion — the mess — feel so risky?

The silence between my question and her response is long and uncomfortable. At last, she begins, "Um … well, in the bathroom, there is usually a mirror, and you know what you look like: tears coming down your face, vomit on your lips and chin. And I have thought, 'If people could see me like this!'" Lisa pauses, then, "What a difference there is between my public and private personas!" She stops again before saying, "I, I always have been insecure. I'm afraid I'm not good enough; I don't measure up. I think bulimia is part of that fear."

*Lisa has stammered through her reply, and I realize that putting the words together has not been easy for her. Yet, in the moments just passed, she has become more visible, more real to me. I have glimpsed her vulnerability, and in the face of that, I have seen my own vulnerability. Slowly, she is coming into focus; I am coming into focus; **we** are coming into focus.*

Lisa's story

When Christine called to make an appointment for our first interview, I felt grateful and excited. Since leaving my hometown, where I often talked with my friend, Elaine, about the eating disorder we both battled, I hadn't found anyone who so fully reflected my struggles.

I hoped Christine and I could open each other's eyes. More than anything, I wanted to see how bulimia had come to live with us and to understand what it would require to show her the door.

I was less sure about what Christine wanted from me. When I arrived at her place, she told me of her interest in how women with eating disorders make sense of their experience. She then spoke about her history as a bulimic woman, and I remember feeling very connected to her.

Time passed quickly. When Christine asked if we could stop, I realized that we hadn't formed any clear insights. Still, just being with her, talking about our clandestine selves, was validating and cathartic.

I left her apartment hoping I'd given her something useful. I respected her as a writer and scholar and wanted her to envision me in a similarly positive light.

Several weeks went by before Christine called again, this time to meet for lunch. I didn't know this would be another interview until I noticed a tape recorder on the table at the German café.

Christine said she had found it challenging to write about our first session. We'd examined bulimia cognitively, she told me, never bringing emotional experience into view.

As a researcher myself, perhaps I should have assumed that Christine was seeking this kind of disclosure from me. Still, I seldom allowed *myself* to peer into the darkest corners of my eating disorder, and I'd never shown that darkness to anyone else, not even to my friend Elaine. I didn't know Christine well; I wasn't sure I could let her in that far. Besides, I didn't hear *her* sharing anything immensely personal. If her doors remained closed, how could I open mine?

And here we were *in public*. Certainly, this woman knew how self-controlled I was; how could she ask for that level of sharing here?

I exposed as much as I could, but it wasn't much. We had our sandwiches and strudel and said goodbye.

Step 6: Christine writes about Lisa

Christine's story

Even with our second interview, Lisa's life history was exceedingly hard to compose. There were several reasons for this. First, I had such varied materials to make meaning from: a journal entry, tapes and transcripts, and fieldnotes from casual conversations. For months, I fretted over what details to highlight and shadow. In addition, unlike the other women's stories, there was no overarching emotional tone to Lisa's account. What a marked contrast between her gut-wrenching journal entry and our detached musings from the first interview!

Another source of turmoil was the fact that, unlike my other participants, Lisa most certainly would read my manuscript. Because of this, when I at last put pen to paper, I wrote through her eyes, feelings, and very soul. I imagined her reaction to every scene, every word. Though I wanted Lisa's story to be instructive to others, in many ways, I composed her chapter *for her*. I needed her to see herself in my work.

As I authored Lisa's story, I witnessed my own story deepening and expanding. Because of this, I decided to include snapshots of my experience alongside hers. In a section titled "Our Mothers," I explored both how Lisa's relationship with her mom may have influenced her approaches to eating and weight and how Lisa's reflections (and my portrayals of those) forced *me* to consider how *my* mother shaped *my* thinking about body and food.

If there was one glaring absence in my chapter on Lisa, it was our fathers. Again and again, as I listened to the tapes and read the transcripts, I encountered her "daddy." Each time, this reawakened my deepest longings for a gentle, affectionate father. To be honest, of all the obstacles to writing Lisa's account, this was the most formidable.

Lisa's story

When Christine began to draft my story, I felt an unexpected apprehension. So often I wondered, "When she looks at me, who does she see?"

I had flashes of panic over this question. During one such flash, I wrote the following passage:

> *This month, Christine sits at her computer and authors a narrative of a woman who struggles with bulimia. The name of that woman will belong to someone else — an anonymous, faceless other — but her story, in essence, will be mine.*
>
> *In my mind, I see Christine's long, graceful fingers tapping away, her burgundy-coated nails creating a light scraping sound as they move across the keys. She pains over the words that describe her life as much as mine.*
>
> *Although I believe in her compassion, I cannot help wondering how Christine will represent me. Will I emerge a protagonist in a battle of crisis and redemption? A weak, helpless woman in a tale of victimization? Will I seem believable? Likeable? Can I face the self I've hidden all these years?*
>
> *Like Christine, I am a writer, and I suppose the task of crafting my story should have been mine. After all, who understands a life better than the one who's lived it? Of course, sometimes others see things that selves cannot — or will not — confront. Might Christine know why I stumbled onto this path? Will she tell me?*
>
> *Sometimes the anxiety overcomes me. Who am I in Christine's pages?*

Step 7: Lisa writes about herself

My apprehension moved me to continue authoring an account for myself. In January 1995, I set out to expand my seven-page journal entry. For weeks, I took time each day to reflect on my relationships with body and food. Using systematic introspection,[11] I closed my eyes, projected myself into various memories, and recorded the images and feelings. Sometimes I smiled at the silliness of my adolescent fixations on fat and weight. Other times, a mass of emotion would rise from my stomach to my throat as I recalled physical pain and emotional loss. From detailed notes, I wrote a series of ethnographic poems and short stories.[37]

Over 100 pages later, I felt drained and frustrated. I still wanted answers to two major questions: "Why do I engage in these self-destructive behaviors?" and, "Why can't I reveal my struggles to my family?" I knew I'd made progress, but I didn't sense completion.

In an advising meeting, Art suggested that I keep searching. "But where?" I asked.

"Maybe you should interview Christine," he said.

Step 8: Lisa interviews Christine

Lisa's story*

I knew there were risks to interviewing Christine: emotional risks associated with peering further into her (and my) eating disorder, and relational risks associated with combining such intensely personal scholarship with an ever-deepening friendship. But the potential for mutual growth as academics, as friends, and as women dimmed the intensity of those risks. I'd already had the unique experience of being a respondent in a qualitative research project; now I could see bulimia, Christine, and myself from the other side of the looking glass.

To help me prepare for our first session in these roles, Christine gave me a 60-minute life history tape she made for a graduate class we both had taken. I'd given Christine mine before her first interview with me. Thinking back, I realized that my tape contained only one passing reference to bulimia — an afterthought, really, as if my eating disorder had stopped by for a brief visit and left, never to be seen again. I wondered what place bulimia would take in Christine's life story.

When I pressed "play," her gentle, soothing voice directed me to a long poem enclosed in a sealed envelope Christine also had provided. I hit "pause," tore open the packet, and pulled out several loose pages. Before reading the title, "Without Daddy," I smiled in anticipation of the childhood I imagined Christine had, a childhood of girlfriend secrets, boy-crazy crushes, and proud-parent events — fonder elements of the childhood *I* remember having.

Eagerly, I pored over each word. But almost immediately, something thick began to seep into the pit of my stomach, and I scanned the pages more and more quickly. "The beatings ... I remember," she wrote of her father. The beatings? My gaze locked on that line for a few moments before I forced myself to continue:

> *They began with the unbuckling ...*
> *how I hated that sound,*
> *   the unbuckling of your belt.*

* Portions of this section are adapted from Reference 20.

> *You'd pull it out of the loops of your pants*
>   *and then fold it over so that it was taut — tight,*
> *so that it would sting,*
> *sting, sting, bare, young skin.*

And it was only the beginning. A cruel father. An unprotective mother. A teenage pregnancy and abortion. Who was this woman I thought I knew? I stared into the corner of the room for 10 minutes before mustering the courage to return to the tape.

Her story jolted me. Depression. Panic disorder. Laxative abuse. I tried to take notes for interview questions, but my eyes couldn't focus on the page. At last, the tape clicked off. Slumping back, I wondered, "What have I gotten myself into?"

Still, I couldn't turn back.

A few days later, I took a deep breath before ringing the bell to her apartment. I'd had some time to muddle through what Christine had shared, and I felt privileged to have been invited inside her times of despair. Given all the emotional and relational turmoil she described, it made sense that Christine would use food to fill the voids.

But when I turned that lens back on myself, I didn't see any such abyss. Having grown up in a stable, loving home, how could I explain my bulimia, to Christine and to myself? Was it a mess I'd created out of nothing?

The door opened, and our brown eyes met. "Hey, Christine," I said.

"Hi! Come in."

We reached the top of the stairs, and she offered an iced tea. As Christine strode to the kitchen, her black dress flowed about her long, graceful legs. I marveled at the beauty and poise undiminished by her often difficult circumstances. She brought two glasses to the living room, and we sat down to talk about school. We spent a few minutes sipping and chatting before I asked if we could begin. Christine nodded, and I turned on the recorder.

I wasn't ready to offer my reactions to her poem and life history tape, so I started with a question that would permit her to talk for some time without interruption. "Christine," I said, "if you were to write the history of your bulimia, what turning points would you include?"

She delved into subjects we'd discussed before, such as obese relatives and her weight-conscious mother, and the experiences Christine delineated resonated strongly with my own. Our conversation flowed comfortably, *too comfortably.*

Christine paused. Then, for perhaps the first time in a face-to-face encounter between us, she ventured into an area where our histories truly diverged. Christine spoke of her pregnancy, quickly and matter-of-factly. She seemed determined, at last, to get it all out. The prospect of opening so fully to each other both excited and frightened me. Part of me wanted to see her more clearly and, in turn, to see myself more clearly. Another part found solace in the blurriness that allowed us to skirt deep emotion. Still, it seemed we'd crossed a threshold.

When Christine finished describing her abortion, which prompted her first eating binge, I asked about another step in her progression toward bulimia. "Can you tell me about the first time you took laxatives?"

"It's still vivid in my mind," she said, and Christine walked me through the process: buying the Ex-Lax, hiding them in a desk drawer, and, eventually, taking a double dose. "Have you ever used laxatives?" she asked.

This caught me off guard. "Never," I answered, not admitting how often I'd considered it.

"I remember blowing out my system that first time," she told me. "You get a lot of cramping. I was sick for a week. Every time I went to the bathroom, I had diarrhea — really intense."

As she spoke, I thought of my own experiences with stomach flu and food poisoning. I recalled the abdominal pressure, the dashes to the bathroom, and the sounds and pungent odors associated with emptying.

I tried to imagine Christine inviting such awful, ugly effects. As we sat in her tasteful, spotless living room, I noted her manicured nails, her shiny, smartly styled hair, and her soft, flawless make-up. In light of all I saw, her purging was almost unthinkable. Then again, I supposed Christine had equal trouble picturing me kneeling in front of a toilet with my finger down my throat.

Her chronology continued. She recalled telling her family of her eating disorder. I wanted to ask Christine how she revealed this to her parents, what words she used, and how they reacted, both at the time and since the disclosure. I ached for her guidance in divulging this to my own parents, who, until 1998, remained unaware of my struggle. This secret haunted me like nothing else in my life. I believed that Christine would be understanding and caring, but my role as interviewer and my ever-present need for self-control prevented me from opening that door.

So we talked about something "safer." She told me of her success with Zoloft®, an anti-depressant. Christine recalled, "I wasn't on it very long before I had no desire whatsoever to overeat, no desire to purge. I felt such an incredible freedom."

But just as a window leading out of bulimia began to appear, she said, "Unfortunately, that didn't last. I took laxatives after I found out a childhood friend had died."

I hesitated, then queried, "Was that the last time?"

"Yes," she responded. "I ruined six months of what an alcoholic would consider sobriety."

I waited for her to inquire about my most recent episode. If she asked, what would I say? Hers happened months ago, mine only days. Christine castigates herself for a rare slip; what would she think about my repeated self-destruction? Was mine a more grave situation than I'd led myself to believe? Christine didn't ask how I was doing, and despite how desperately I longed to share my fears, I couldn't quite bring myself to do it.

Again we returned to a comfortable place of connection, this time, around our mothers. Fighting the urge to remain in that buffer zone, I summoned the strength to inquire about the disclosure. "Wh — what has been … your parents' response to your eating disorder?"

Christine stared at me intently, as if unsure how to proceed. Slowly, she began, "My mother and I were sitting together on the front porch. I wanted her to read some narrative vignettes I'd composed about my struggles with bulimia. This was really important to me. I hoped it would bring us closer somehow. After finishing the manuscript, my mother closed it and said, 'You're too smart for that — far too intelligent to do that to your body.' Then, she *chuckled*, looked away and added, 'Goodness, you always had such a wonderful imagination!' My mother was — and is — in denial about the whole thing."

Wham! Christine's experience added yet another "worst case scenario" to my mind's already-vast repertoire.

A storm of emotion twisted and swirled inside me. Though terrifying, there was one more door I had to open. Almost breathlessly, I said, "Christine, before I read your poem and heard your life history tape, I assumed that you and I were more similar in our experiences. I want to share the last part of a reaction I wrote after your tape clicked off":

> *Something overcame me as I read and listened to Christine's life history, something unexpected and frightening. Bulimia, in the context of Christine's life, made perfect sense to me. She used addictions as means of coping with her disordered life. I responded to her account by thinking, "Ah, yes. So **that's** why." At the same time, my own story began to make less and less sense. What reasons could I offer to explain my bulimia? What empty spaces?*

Putting down the paper, I cleared my throat and pushed on. "I guess what I'm saying is this: confronting your story, I no longer feel like *I* have a plausible account for my eating disorder." My eyes filled with tears, but I blinked them away.

Silence engulfed us. Finally, she said, "I told you before that yours was the hardest story for me to write. I said it was because I had trouble tapping into the emotionality of your experience. But it was more than that. In fact, the biggest stumbling block for me was your relationship with your father." She paused, searching for words. "Listening to you talk about your dad — during the interviews, and later on the tapes — was almost too much for me to bear. I so wanted to *be you* in those stories. You had a daddy."

Her words stunned me, and I spent a few seconds collecting my thoughts. "Yes," I began tentatively, "I suppose that's true, but my relationship with my father also is somewhat one-sided. I find it very easy to share with him the brighter moments of my life and quite difficult to expose the darker ones. Your father may not have been a daddy, Christine, but he must have opened a space where you felt permitted to reveal your mess. I'm still hiding mine."

She processed what I'd said. "Having to reconfront the difference between your father and my father was what scared me the most about our session today. Your story of your father is the story that I want — that I've *always* wanted — for myself. But now I'm getting a different view."

In that moment, I felt connected to her like never before. A small but significant piece of our collective picture had come into focus. We talked on for several minutes until I asked, "Are you hungry, Christine?"

"Starved. Should we get lunch?"

"Absolutely," I said. The food that afternoon tasted better than any I'd had in recent memory. For the first time in months, both my stomach and my heart felt full.

Christine's story

As the date for our "turning the tables" interview drew closer and closer, I found myself growing more and more anxious. What would I be asked to reveal? What would Lisa reveal? What might I learn about myself, about Lisa, about *us*?

At the same time, I experienced a firm resolve. I committed myself to unmasking "it" (although I wasn't sure exactly what "it" was). I would hold back nothing.

A week or so before we met, I gave Lisa a life history tape and a poem I'd written about my father. These materials detailed the most raw moments of my experience. They contained my mess — the injustice, ugliness, and pain in my life — and they shouted, "This is who I am!" On tape and in print, I uncovered myself without shame. I realize now that saying those things in an interview would have been nearly impossible; I guess taping and writing them in private felt safer somehow.

Though they probably overwhelmed Lisa, the tape and poem set the agenda for the first part of our session. She began by asking me to elaborate on passages I'd composed about my pregnancy and abortion and to explore their connection to my bulimia. While responding to her requests, I was astounded by how unemotional — even monotonous — my voice sounded. I felt detached, almost as if I were talking about someone else. At times, I even *laughed* about my struggles. My reactions disturbed me. I was amazed at how alienated I was from my experience. Whenever I heard myself speak, I felt sad for the woman I was describing (me). How I wished I could cry for her (me).

But it was Lisa who almost came to tears. After she read her reaction to my life history tape and poem, a long silence ensued. I could sense how deeply troubled she was.

Since the start of this process, bulimia was a main point of connection between us. Perhaps both of us had been blinded by our similarities. Now came the real test — seeing what could be learned through our differences. I wanted to say something comforting, to communicate that bulimia is a complex response to complex circumstances, and those circumstances will be different for everyone. I chose another course instead.

To show Lisa that I valued the risk she took in sharing her written response with me, I decided to lift the remaining veil between us. I said that I had longed for us to connect on the issue of our fathers and felt profoundly isolated when we didn't. By this, Lisa seemed as taken aback as I had been by her reaction to my life history materials. But then she told me of *her* profound isolation. Lisa said she desperately wished for the courage to be "real" with her father and with others in her life.

Through tears, we exchanged an intense, sisterly glance. She may not have realized it, but I knew that Lisa had just demonstrated the very courage she prayed for. We both had. It was a new beginning.

Step 9: Christine reads Lisa's autoethnography

Not long after that interview, Lisa gave me a copy of a paper she wrote about her history as a bulimic woman.[37] I tore through each page, eagerly consuming every word. As she had in her journal entry, Lisa graphically represented her torment and suffering.

The manuscript portrayed the messiness in Lisa's life. In a poem, she highlighted the enormous pressure she felt to be perfect. The honest details Lisa presented showed me that her life and relationships were less than idyllic. Seeing her imperfection drew me closer to her than ever.

Another important aspect of this piece was that at every turn, Lisa's storied experience evoked memories of my lived experience. The layers of her struggles moved me to think more carefully, to feel more deeply, and to write more purposefully about my own struggles.

Step 10: Lisa writes about Christine

Lisa's story

Following another interactive session, I immersed myself in the details of Christine's life. Striving to do justice to her story, I took on her experiences as my own. I listened repeatedly to Christine's voice on the interview and life history tapes. Then I transcribed the tapes and studied the transcripts many times. After several weeks of this, I felt "inside" her life, her body.

I wanted my renderings to reflect the emotionality of Christine's experience (and the emotionality of *my* experience of her experience). Because of this, I chose a form — the ethnographic short story — that could express her trials evocatively.

But then I wondered about the consequences of expression *for Christine*. More than my research "subject," this beautiful, articulate woman had become my friend. As her confidant, I wanted so much to shield her from her pain. She'd lived her father's abuse and her abortion; how could I ask her to confront the darkness again? Christine told me that she couldn't imagine her story without these events, so I featured them. But I cried every step of the way, at times having to stop and catch my breath. I saw the images, felt the feelings, and painted the scenes. Her experiences broke my heart at times, as if they had happened to me, but I wrote through them and moved Christine's character, as Christine herself had moved, toward a place of healing.

To gain additional detail and insight, I interviewed her a third time. After transcribing our conversation, I returned to the narratives, making appropriate additions and deletions. Based on what she shared, I often found myself adding to the horror and deleting my own wishful projections. I did save a ray of hope, however — the ending, an ending that points to a future of self-acceptance. I could not change the events of Christine's life, but I could direct her toward something else. She deserved no less.

Christine's story

I nervously anticipated Lisa's final draft — her story of my story. I empathized strongly with her. I knew how hard it was for her to write about my life, because I knew how hard it had been to write about *her* life.

In the weeks that passed, numerous questions filled my mind. How would she cast the "characters," including me? What episodes and storylines would she spotlight? Would my experience "warrant" an eating disorder? Most importantly, would Lisa make my bulimia make sense?

Step 11: Lisa reads Christine's dissertation

One morning in November 1995, I noticed a thick document in my mailbox at school. I knew immediately what it was — Christine's completed dissertation.[25] Just seeing it sent my heart racing, and I was overcome with excitement, curiosity, and fear. Because I was still finishing my piece on her, I tucked the bound copy away for a week or so.

The same day I completed my draft of Christine's story, I sank into my reading chair, taking in every line of her dissertation, from preface to references, in one sitting. I resisted the impulse to skip to the chapter about me because I didn't want to miss anything. Two hours later, I reached the section about "Eileen," my pseudonym, and my breathing became short and shallow. I realized that here, in black and white, would be some of the most personal and wretched details of my life.

As I scoured each page, Christine's story of Eileen forced me to face both the implications of my deception and the darkest moments of my self-loathing. All the while, though, I was overcome with an amazing calm.

When I finished at 3:00 a.m., I felt *relieved*, so relieved. There it was — an account of my experience with a beginning, middle, and end. I liked this character Eileen. Though self-critical and obsessive, she was strong and smart. Her life — and even her bulimia — made sense. Revealing myself to Christine (and confronting Eileen, my sister self) had been difficult. But beneath the masks of my most painful secrets, Christine — my friend, my ethnographer, my *inter-viewer* — had found a competent human being. Perhaps if she saw me that way, I could too. I slept well that night.

Step 12: Christine reads Lisa's narrative ethnography

My heart skipped a beat and my hands shook the second I lifted Lisa's narrative ethnography[36] out of my mailbox at school. I was on my way to give an exam to my Communication Theory class.

The students began writing, and I settled in for the quiet two hours I would have to read the document. But almost immediately, I started crying. The images were too much. I closed the paper.

When I returned home that night, I read on. All the while, I thought, "My God, is this really me?" I cried for whomever that paper was about. I guess I cried for myself, although it didn't feel that way. I read and reread the manuscript several times. Each time, I turned away with that same disbelief.

Lisa's ethnography showed me the underlying meanings of my bulimia. Her emphasis on my family background and my complex adolescence affirmed that my unhealthy relationships with body and food had more to do with deep emotional losses and voids than with a quest for thinness. I so appreciated the depths to which Lisa thought about the functionality of my eating disorder. Her work taught me to look more deeply into the lives and stories of anorexic and bulimic women, and to look more deeply into my life and story. Our project had come full circle.

Commentary

With our co-narrated confessional tale, we attempted to show how we learned to see each other, ourselves, and the layers of complex emotionality associated with eating disorders through the reflexive lenses of researcher, respondent, auto- and narrative ethnographic writer, and reader. In the discussion that follows, we review key insights we reached through each of these lenses.

Lessons from researchers

During our tenure as researchers, each of us deepened and expanded not only her comprehension of the respondent's experience but also her own story of bulimia. In many cases, this happened because stories shared by the respondent held up new and unexpected mirrors to the researcher's experience.

For example, Lisa's talk of a self-critical mother moved Christine to reflect on, and later to write about, her own mother's body image and how this may have influenced the ways Christine related to her body and to food. For Lisa, Christine's descriptions of her self-destructive adolescence so mirrored her past that she began to examine more closely the larger contexts (e.g., the peer culture) in which her bulimia emerged.

While we learned valuable lessons from such similarities, even greater levels of understanding were reached when confronting our differences. But those moments were far from comfortable.

Our vulnerability is perhaps most clear in our portrayals of the dissonance each of us experienced when encountering particular chapters in her respondent's story. Christine's strongest discord involved the loving relationship Lisa reported having with her father. Christine, who had come to think of fathers as antagonists in bulimic women's accounts (including her own), faced a sense-making crisis. She didn't understand how Lisa could have both a "daddy" and an eating disorder. At first, Christine heard Lisa's story of her father as a story she always had wanted for herself. Christine's renewed grief over her troubled relationship with her father began to subside only when Lisa disclosed how difficult it was for her to expose imperfections to her father.

In her stint as researcher, Lisa experienced dissonance as she listened to Christine's life history tape, which chronicled a series of losses and voids Lisa thought someone might, rather rationally, try to fill with food. But for Lisa, the "logic" of Christine's story called her own account into question. "What reasons could I offer to explain my bulimia?" Lisa wondered. "What empty spaces?"

By exploring our similarities and differences, we have come to understand that, at various points in the research process, respondents' stories are reflections, deflections, extensions, and/or projections of the researcher's own story. As we have seen, accounts offered by those we study can plunge us into our own despair, and they can muddy the waters of our life stories. It is therefore an act of courage to take in another's trials, to make meaning from them, and to use them as gifts of comfort, companionship, and inspiration, in our fieldwork reports and in our own lives.[7]

With this act of courage, however, comes a delicate balance between the contradictory demands for expression and protection. On one hand, research moves us to know and understand both others and ourselves. But when we research emotional topics and we commit ourselves to field relationships, we reach depths of experience that trigger our defenses, which may lead us to shield self and others from pain.

Christine, for example, initially felt compelled to conceal certain parts of her story from Lisa, the respondent who was most fully her peer. Christine was anxious about how these aspects would affect Lisa's perceptions of her, about what disclosures *her own* disclosures might prompt from Lisa, and about how Lisa's reciprocal disclosures might then alter Christine's story of herself.

The same kinds of fears surfaced when Lisa assumed the researcher role. Lisa was reluctant to ask Christine how she told her parents of her bulimia and what their responses were. Lisa knew that, in order to understand Christine's account of her hidden identity, she had to learn how Christine managed secrecy and disclosure. But Lisa hesitated at first because she believed that Christine's experience could preview her own. Lisa reasoned that if Christine's parents dismissed her plight or rejected their daughter, perhaps hers would too, and few prospects frightened Lisa more than those.

As protective as each of us sometimes was of herself, she was no less protective of her respondent. Early on, Christine recognized that much of Lisa's vulnerability centered on her reluctance to share her eating disorder with significant others, so Christine took extra care when asking Lisa to venture into this area. Lisa demonstrated equal caution when she requested that Christine talk about her father's abuse and her abortion.

That we exhibited some of our greatest sensitivity in those areas where our histories most diverged confirms for us that, when studying emotional experience, it is not necessary for researchers to have personal experience with the topic under investigation. To understand the emotional experience of bulimia, for example, it seems less important that researchers have a history of engaging in bulimic behaviors than that they can commit themselves to "taking on the lives" of their respondents.

On another practical note, because research on emotional experience demands this kind of circumspection, we believe that a deep understanding of another's struggles requires more than a single interview. Each of us had two in-depth sessions with the other, and Lisa interviewed Christine a third time after completing a draft of her story. We also met informally several times for coffee or lunch. These meetings were important because they lacked the pressure of the interview situation. To other qualitative researchers, we strongly recommend alternating formal, recorded sessions with everyday shared activities.

In short, our time spent as researchers taught us that, when studying emotional topics, we become what Behar[3] calls "vulnerable observers." By confronting the joys and horrors of others' experience, we face the joys and horrors of *our* experience. Because of this, we must ask ourselves before embarking on such a project: am I prepared to take on another's full humanity and to explore and unveil my own?

Lessons from respondents

We recognize that not every researcher will have the opportunity or desire to turn the tables. For us, though, the experience was highly instructive.

As qualitative researchers, we spend a lot of time worrying about the harms that may come to our participants. We fret over emotional risks, confidentiality, and informed consent. But as respondents, we saw as never before the *benefits* of participation, such as human connection, catharsis, and self-understanding.

The interview sessions provided each of us with unique opportunities to confide in someone knowledgeable, interested, and caring. For Lisa, Christine was the first person she felt she could "really talk to" about her bulimia since leaving her hometown in 1989. For each of us, it was comforting and validating to have someone in her life who was trying, with every ounce of her energy, to understand her experience.

The presence of an invested other permitted each of us to work through and express emotions. Uncovering our most closely guarded secrets was frightening at first, but ultimately, quite empowering. In our time spent as informants, each of us released some of the guilt associated with her (once-) concealed self.

The respondent lens also showed us that qualitative research on emotional experience involves more than just the collection of meaningful accounts; it also helps participants *construct* such accounts. As a participant in Christine's study, Lisa came to make sense of her bulimia as a response to perfectionist standards and her failure to achieve them. As Lisa's informant, Christine interpreted her eating disorder less as a program of weight control (as she had perceived it when composing "Writing it Down") and increasingly as a means of coping with personal and relational losses. For both of us, our tenure as respondents gave us a much more profound understanding of eating disorders in general, and our own bulimia in particular.

At the same time, however, our experience revealed that participants might be unable to tap into the depths of their emotional experience until a caring and trusting relationship develops with the researcher. This requires more than just "rapport." During Christine's first interview of Lisa, Christine proved quite adept at creating a context in which Lisa was comfortable talking about a host of issues she rarely discussed. Nevertheless, even in our second session, Lisa felt unprepared to examine her darkest moments — let alone to expose them. In fact, it was a full year into our relationship before both of us were ready to "get it all out." The long-term nature of this type of project is both one of its greatest strengths and its most practical drawback (particularly for researchers under pressure to graduate and/or publish in a timely fashion).

Our research also taught us the importance of finding an appropriate setting for emotional exposure. Christine conducted her second interview of Lisa in a café. While this was intended to naturalize their interaction, and the session did produce meaningful insights into Lisa's struggles, Lisa found the space too public for the kind of disclosure Christine was seeking. All other sessions were held at Christine's apartment. As a respondent, Lisa felt much more at ease there. In addition, the arrangement, décor, and artifacts of Christine's place gave Lisa a better sense of the kind of person Christine was (e.g., organized, tasteful, and concerned about appearances). When the tables turned, Christine wanted to stay on her home "turf" because this gave her a sense of comfort and control. This could have worked against disclosure because Lisa's self-control was exactly what had obstructed Christine's understanding of her. As it turned out, our highest level of sharing took place during Lisa's first interview of Christine. Perhaps it was the combination of the right time (a year into our relationship) and the right setting (Christine's apartment).

Of course, the right setting will vary from project to project, person to person, and session to session. Because different sites offer advantages and disadvantages, perhaps the best solution would be to alternate interview sessions (and perhaps more casual encounters as well) between the researcher's preferred setting and the respondent's preferred setting.

Nonetheless, our experience leaves us suspecting that, for some respondents, the right time never comes, and/or the right place is never found. We pushed each other and ourselves so hard in part because we both are academics who would have felt empowered to call off the project if it became emotion-ally overwhelming. We are concerned that non-academics would not feel so empowered, and this means that those who study emotional topics contin-ually must question which doors we rightfully can ask participants to open and which are better left closed.

It might help other qualitative researchers to know that, during our stints as respondents, there were doors each wanted (and was willing) to open, but *not always in the presence of the fieldworker.* For us, some emotional expe-riences were shared more easily when recording and writing independently than when being interviewed. Although we are reflective and articulate about our struggles, each of us deepened her researcher's understanding of her experience by sharing a life history tape and autoethnographic texts.

Each of us had taken a graduate course in which students were required to narrate and record their life stories. To help generate questions for her first session with Lisa, Christine asked for Lisa's life history tape, and when their roles reversed, Christine gave her tape to Lisa. These tapes provided insights that might not have been shared in an interview (or may have required several more sessions to reach).

Listening to Lisa's tape, Christine noted how peripheral bulimia seemed to her life story; the subject merited only a passing reference. This foreshad-owed the difficulty Christine later would face in trying to access the "mess" beneath Lisa's public persona. When the tables turned, Lisa was shocked by some of Christine's recorded disclosures, especially those related to her father's abuse and her abortion. However, encountering these disclosures this way allowed Lisa to work through some of her thoughts and feelings before the interview.

Written accounts added yet another layer. Each of us found herself able to share things in print that she would have been reluctant to reveal in a face-to-face encounter. For example, on paper, each of us could describe vividly the embarrassing details of our purges; while being interviewed, we only hinted.

Our experience suggests that alternative means of accessing emotional experience can add breadth and depth to qualitative projects. Not all partic-ipants will feel comfortable speaking their life stories onto tapes, and not all will have practiced autoethnographic writing. Still, when encountering a seemingly locked door, researchers might consider asking respondents to keep a journal, to take or share photographs, to compose short stories or poetry, to paint, or to do anything else that might offer a more complete understanding of their emotional experience.

Lessons from writers

In this project, we engaged in what Carolyn Ellis[10] calls "emotional sociol-ogy." That is, we researched a topic that was emotional for us, and we conducted our study as much with our hearts as with our heads. The next step was to write texts that featured and evoked emotional experience.

In recent years, some qualitative writers have accomplished this by including powerfully emotive quotes.[23,29,39] Others have composed evocative narrative accounts from fieldnotes and interview transcripts.[5]

Following their leads, Christine wrote her chapter on Lisa as a life history that brings together auto- and narrative ethnographic poetry and prose. Later, Lisa's account of Christine took shape as a series of ethnographic short stories.

By featuring our own and each other's emotional experience, we were writing for nontraditional readers. Our goal was to create works that engaged our audience not only intellectually but also emotionally, ethically, and aesthetically. While writing, we considered our readers not as passive receivers of definitive conclusions but as co-participants in ongoing conversations about the lived experiences of bulimia and fieldwork.

To compose such a text, qualitative writers must get inside their respondents' (and their own) emotional experience. Although time-consuming and tedious, doing our own transcribing was invaluable, as was listening repeatedly to interview tapes. For each of us, these practices gave a sense of the other's voice and speech patterns. By the time each sat down to write, she almost could speak (and write) as the other. So much intersubjectivity was achieved that sometimes, when looking over collaborative work from 1995 and 1996, we can't tell who the author is. We often ask each other, "Did I write that about myself, or did you write that about me?"

In addition, our tenure as auto- and narrative ethnographic writers underscored the importance of considering participants an audience.[13] Feeling emotionally and morally compelled to "do it right" and *do right by* each other, each wrote both about and *for* her respondent.[21] Often, each of us felt protective of the other, hoping our renderings would offer her tools for coping and not further pain.

Finally, we learned that composing an account of another's emotional experience can break one's heart.[3] So immersed in the other's life world, each of us wrote "from the gut," through fury, pain, and ambivalence. Each of us cried as she wrote about the other, feeling compassion for her in ways she often couldn't feel for herself.

Lessons from readers

Traditionally, it has been assumed that research participants care little about — and won't read — fieldwork reports.[13] In our project, however, each of us felt *deeply* invested in how the other represented her experience. We cannot say how much of this is due to the fact that both of us are researchers and writers. What we can say is that each anxiously awaited the other's renderings.

Just seeing the documents — Christine's completed dissertation and Lisa's article-length narrative ethnography — was jarring. For each of us, this moment previewed the powerful experiences she would have reading it.

Taking in the other's images and words frequently was painful. The written accounts forced each of us to confront her darkest moments of binging, purging, secrecy, and betrayal. When reading Christine's manuscript, Lisa had to face the chasm between her public persona and the private war she waged against bulimia. For Christine, seeing Lisa's constructions of her father's abuse and her abortion called up long-suppressed grief.

On the other side of such pain, however, were many gifts. In each other's work, we found accounts that rendered even our most unintelligible experiences intelligible. For years, each of us had struggled to make sense of her eating disorder. Our cycles of binging and purging often baffled and frightened us. We would wonder, as someone without bulimia might, "Why would anyone do this to herself?"

In her dissertation, Christine theorized that Eileen's (Lisa's) bulimia was rooted in perfectionism. That is, Eileen set impossibly high standards, and when she failed to achieve them, she would feel guilty and punish herself. Both the unreasonable demands and the guilt, Christine suggested, could trigger bulimic behaviors. Lisa was relieved to see that, in Christine's renderings, her eating disorder wasn't, as Lisa had feared, "a mess I'd created out of nothing." Christine found Lisa's narrative ethnography similarly instructive. Lisa's portrayals moved Christine further beyond a superficial understanding of her bulimia. As she read, Christine saw more clearly than ever before that her eating disorder had more to do with personal and relational voids than with a desire to be thin.

As readers, each of us found additional confirmation in the fact that the other constructed her as a protagonist who was capable of moving past bulimia. In Eileen, Lisa saw much more than "just a bulimic." She saw an intelligent and determined woman. Christine's dissertation helped Lisa realize that, if Eileen (Lisa) could learn to accept her human imperfections, she would have little "need" for an eating disorder. For Christine, Lisa's manuscript showed her how much she *already* had overcome, and it gave her the hope that, if she found healthy ways to "feel full," the impulse to engage in bulimic behaviors likely would subside.

The pages of each other's work also invited us to feel for ourselves. Through the other's words, each grieved her losses, cringed at her obsessions and, occasionally, even smiled at herself. For Christine in particular, the reading experience allowed her to reconnect with a history of emotional experience from which she usually felt detached.

Often, reading what the other wrote triggered memories that moved us to write even more about ourselves. Thus would begin another cycle of writing, exchanging, reading, and talking.

In short, having the opportunity to read another's reflections on our emotional experience added depth and dimension to our stories. This reinforced for us the importance of taking this kind of scholarship back to respondents for examination, critique, and further dialogue. We believe that such texts can be as beneficial to our participants as to our academic colleagues (and to our own careers). But this only can happen if we give respondents the chance to become *reader* respondents.

Conclusion

In the course of this project, we witnessed intersections and divergences between the researcher's (spoken and written) stories of her emotional experience, the respondent's (spoken and written) stories of her emotional experience, and the researcher's stories of the respondent's experience. In unexpected and profound ways, each of these stories impacted and was impacted by every other. Life stories and research stories perpetually shaped and reshaped one another.

This chapter comes to a close, but our identity-transforming and emotional project goes on. We continue to share stories, to fashion and refashion each other and ourselves.

> *Every step of the way,*
> *each of us holds a mirror for the other,*
> *a mirror that can reveal her to be*
> *a beautiful,*
> *competent human being,*
> *if only she will*
> *stop for a moment*
> *and look.*
>
> —by Lisa Tillmann-Healy.

Acknowledgment

The authors wish to thank Kathleen Gilbert for her encouragement and guidance and Carolyn Ellis and Art Bochner for their unwavering support, both professional and personal. Anyone with questions about or reactions to this project should contact Lisa Tillmann-Healy at Rollins College, 1000 Holt Ave. Box 2723, Winter Park, FL 32789 (or ltillmann@rollins.edu).

References

1. Angrosino, M., *Opportunity House: Ethnographic Stories of Mental Retardation*, AltaMira Press, Walnut Creek, CA, 1998.
2. Austin, D.A., Kaleidoscope: The same and different, in *Composing Ethnography: Alternative Forms of Qualitative Writing*, Ellis, C. and Bochner, A.P., Eds., Alta-Mira Press, Walnut Creek, CA, 1996, 206–230.
3. Behar, R., *The Vulnerable Observer: Anthropology that Breaks Your Heart*, Beacon Press, Boston, 1996.
4. Bochner, A.P., Perspectives on inquiry II: Theories and stories, in *Handbook of Interpersonal Communication*, Knapp, M.L. and Miller, G.R., Eds., Sage, Thousand Oaks, CA, 1994.
5. Cherry, K., Ain't no grave deep enough, *J. Contemporary Ethnography*, 25, 22–57, 1996.

6. Conquergood, D., Rethinking ethnography: Towards a critical cultural politics, *Commun. Monogr.*, 58, 179–194, 1991.

7. Coles, R., *The Call of Stories: Teaching and the Moral Imagination*, Houghton Mifflin, Boston, 1989.

8. Denzin, N.K., *Interpretive Ethnography: Ethnographic Practices for the 21st Century*, Sage, Thousand Oaks, CA, 1997.

9. Denzin, N.K. and Lincoln, Y.S., Eds., Introduction: Entering the field of qualitative research, in *Handbook of Qualitative Research*, Sage, Thousand Oaks, CA, 1994, 1–17.

10. Ellis, C., Emotional sociology, in *Studies in Symbolic Interaction*, Denzin, N.K., Ed., JAI, Greenwich, CT, 1991, 123–145.

11. Ellis, C., Sociological introspection and emotional experience, *Symb. Interact.*, 14, 23, 1991.

12. Ellis, C., "There are survivors": Telling a story of sudden death, *Sociolog. Quarterly*, 34, 711–730, 1993.

13. Ellis, C., Emotional and ethical quagmires in returning to the field, *J. Contemporary Ethnography*, 24, 68–98, 1995.

14. Ellis, C., *Final Negotiations: A Story of Love, Loss, and Chronic Illness*, Temple University Press, Philadelphia, 1995.

15. Ellis, C., Speaking of dying: An ethnographic short story, *Symb. Interact.*, 18, 73–81, 1995.

16. Ellis, C. and Bochner, A.P., Telling and performing personal stories: The constraints of choice in abortion, in *Investigating Subjectivity: Research on Lived Experience*, Ellis, C. and Flaherty, M.G., Eds., Sage, Newbury Park, CA, 1992, 78–101.

17. Ellis, C. and Bochner, A.P., Eds., *Composing Ethnography: Alternative Forms of Qualitative Writing*, AltaMira Press, Walnut Creek, CA, 1996, 13.

18. Ellis, C. and Bochner, A.P, Introduction: Talking over ethnography, in *Composing Ethnography: Alternative Forms of Qualitative Writing*, Ellis, C. and Bochner, A.P., Eds., AltaMira Press, Walnut Creek, CA, 1996, 13–45.

19. Ellis, C. and Bochner, A.P., Autoethnography, personal narrative, reflexivity: Researcher as subject, in *Handbook of Qualitative Research*, 2nd ed., N.K. Denzin and Y.S. Lincoln, Eds., Sage, Thousand Oaks, CA, 2000.

20. Ellis, C., Kiesinger, C.E., and Tillmann-Healy, L.M., Interactive interviewing: Talking about emotional experience, in *Reflexivity and Voice*, Hertz, R., Ed., Sage, Thousand Oaks, CA, 1997, 119–149.

21. Fine, M., Working the hyphens: Reinventing self and other in qualitative research, in *Handbook of Qualitative Research*, Denzin, N.K. and Lincoln, Y.S., Eds., Sage, Thousand Oaks, CA, 1994, 70–82.

22. Fisher, W.R., Narrative as a human communication paradigm: The case of public moral argument, *Commun. Monogr.*, 51, 1–22, 1984.

23. Greenspan, H., Lives as texts: Symptoms as modes of recounting in the life histories of holocaust survivors, in *Storied Lives: The Cultural Politics of Self-Understanding*, Rosenwald, G.C. and Ochberg, R.L., Eds., Yale University Press, New Haven, CT, 1992, 145–164.

24. Jackson, M., *Paths Toward a Clearing: Radical Empiricism and Ethnographic Inquiry*, Indiana University Press, Bloomington, 1989.

25. Kiesinger, C.E., Anorexic and Bulimic Lives: Making Sense of Food and Eating, unpublished doctoral dissertation, University of South Florida, Tampa, 1995.

26. Kiesinger, C.E. and Kiesinger, J., *Writing it Down: Sisters, Food, Eating, and Our Bodies,* Unpublished manuscript, University of South Florida, Tampa, 1992.

27. Punch, M., Politics and ethics in qualitative research, in *Handbook of Qualitative Research,* Denzin, N.K. and Lincoln, Y.S., Eds., Sage, Thousand Oaks, CA, 1994, 83–97.

28. Richardson, L., The consequences of poetic representation: Writing the other, rewriting the self, in *Investigating Subjectivity: Research on Lived Experience,* Ellis, C. and Flaherty, M.G., Eds., Sage, Newbury Park, CA, 1992, 125–137.

29. Riessman, C.K., Making sense of marital violence: One woman's narrative, in *Storied Lives: The Cultural Politics of Self-Understanding,* Rosenwald, G.C. and Ochberg, R.L., Eds., Yale University Press, New Haven, CT, 1992, 231–248.

30. Ronai, C.R., Multiple reflections of child sex abuse: An argument for a layered account, *J. Contemp. Ethnography,* 23, 395–426, 1995.

31. Rosenwald, G.C. and Ochberg, R.L., Eds., *Storied Lives: The Cultural Politics of Self-Understanding,* Yale University Press, New Haven, CT, 1992.

32. Schwalbe, M., The mirrors in men's faces, *J. Contemp. Ethnography,* 25, 58–82, 1996.

33. Spence, D., *Narrative Truth and Historical Truth,* W. W. Norton, New York, 1982.

34. Stoller, P., *The Taste of Ethnographic Things: The Senses in Anthropology,* University of Pennsylvania Press, Philadelphia, 1989.

35. Tedlock, B., From participant observation to the observation of participation: The emergence of narrative ethnography, *J. Anthropol. Res.,* 41, 69, 1991.

36. Tillmann-Healy, L.M., *Bulimic Biographies: Through a Shared Silence,* unpublished manuscript, University of South Florida, Tampa, 1996.

37. Tillmann-Healy, L.M., A secret life in a culture of thinness: Reflections on body, food, and bulimia, in *Composing Ethnography: Alternative Forms of Qualitative Writing,* Ellis, C. and Bochner, A.P., Eds., AltaMira Press, Walnut Creek, CA, 1996, 76–108.

38. Van Maanen, J., *Tales of the Field: On Writing Ethnography,* University of Chicago Press, Chicago, 1988.

39. Wiersma, J., Karen, The transforming story, in *Storied Lives: The Cultural Politics of Self-Understanding,* Rosenwald, G.C. and Ochberg, R.L., Eds., Yale University Press, New Haven, CT, 1992, 195–213.

section two

A broader perspective

chapter six

Qualitative research as a spiritual experience

Paul C. Rosenblatt

Contents

Introduction

I have a story to tell you, about a spiritual journey, a journey I didn't know I was on and didn't intend to be on. It's about how spiritually remarkable things can happen to one when one carries out qualitative research.

Years ago I thought personal stories were useless. I didn't think data from a single case proved anything; there were too many ways to understand a single case. Also, I didn't trust self reports, because I thought they were not objective and that their biases would be self-serving. I still have my doubts about a single case, but there has been an enormous upheaval in the social sciences — one that has changed the way that many of us think about everything, including personal narratives. The upheaval has led to radical

0-8493-2075-5/01/$0.00+$.50
© 2001 by CRC Press LLC

new approaches to carrying out, writing, understanding, and communicating research. And it has led to critical revision of how to evaluate the
approaches and claims of what could be called "traditional" approaches in
the social and behavioral sciences.

The changes have come from many different sources. Prominent among
them have been developments in postmodernism,[16] social constructionism,[12,14]
feminism,[2,11] qualitative social research,[9] discourse analysis,[4,13] the critical social
sciences,[8] and thinking about race and ethnicity.[1] For many of us, these perspectives have changed everything, including our attitudes toward whose
voice should be heard in reports of research, what is knowledge, what is truth,
what are the proper uses of social research, and the biases in supposedly
objective positivist, quantitative research.[3]

In the process of changing how I think about everything connected to
research on human social life I have come to value personal narratives. I still
see the potential biases in narratives, and I still can find value in what some
might call the "scientific" approaches to understanding people. But I see the
biases and limitations of the "scientific" approaches. At the same time I see
the strengths in all sorts of other approaches, including narratives. The
strengths of narratives include the details, the information about contexts, the
power of a connected story line, the openness and clarity about meanings, the
depth of feeling, and the modesty of theoretical claims. I don't need to take
sides, but in a world where many people take sides and say that quantitative,
positivist, modern social research is best — far superior to an isolated personal
narrative or even to research based on a substantial number of personal narratives — I am clear that from some perspectives they are wrong about their
own approaches to research and wrong about the value of narrative.

My more recent qualitative research has reflected the importance I give
to narratives. In much of that research I have tried to understand family life
by pulling together the information from narratives — for example, narratives I have generated through interviewing individuals or families or the
narratives contained in personal diaries. Even as I try to generalize from a
set of narratives, I highlight the diversity and details in those narratives —
for example, the narratives of bereaved parents[22] or of members of farm
families caught up in the powerful economic forces that have forced millions
of American farm families out of farming.[19] I would prefer to offer such work
here, pulling together the stories of a number of qualitative researchers who
have been affected spiritually by their research experiences. But I don't have
such research to offer. And in some ways I think focusing on one person's
narrative is superior — because it is a more coherent, more context-sensitive
approach. Anyhow, if I were going to do research on qualitative researchers'
spirituality, I almost certainly would begin by doing what I do in this chapter:
getting clear about my own story. Not infrequently, qualitative research
begins with a single story, and sometimes it's the researcher's. I offer my
story to you. Perhaps you can develop your own research project, taking off
from this first story, and perhaps the chapter will sensitize you to where you
have been and may go as a qualitative researcher.

Narrative of a spiritual journey

Imagine a religion in which a major sacrament is to listen to strangers talk for several hours about the deepest, hardest, most painful, most challenging experiences of their lives. For me, qualitative interviewing has been such an experience. Years of qualitative research on the difficult issues in people's lives have had a profound impact on me spiritually, although I started out having no idea that would happen or even that spirituality would be relevant to my research.

I hesitate to define spirituality, because I know the experience and understanding of it are different from person to person and even different within a person from time to time. Also, I think some of what I experience as spirituality cannot be captured in words — or maybe it's just that the words make what I have experienced seem more ordinary, definite, and cognitive than I experienced them. The kinds of things I think of as spirituality include a sense of what is most important in life, what it means to be human, the most fundamental meanings of connections with other humans and with the universe, the meanings of life and death, and what is felt and known about certain things that are at the edge of or beyond rationality and immediate experience. I also think of spirituality as including the ways in which we are in a oneness with others, awe as an experience and what causes feelings of awe, the ways we can find profoundly moving and transcendent experience in what might seem to many people to be ordinary, a sense of basic connection to lives and times in the past and in the future, and a sense of relationship with God, however one defines God.

For me, qualitative research has been a spiritual journey. In some ways that has been wonderful. In some ways it has been painful, disquieting, identity-disrupting, confusing, frightening, and relationship-rocking. Looking back on my good days, I can believe that those difficult feelings and experiences have probably been essential to a spiritual transformation process that I am glad I have been going through. But there were experiences that I would have gladly done without, whatever the possibility of transformation that might come with those experiences. Thus I write this partly as an invitation to a spiritual experience and partly as a warning (because spirituality is something like nuclear fission in that it can provide energy to heat and light your home or make a very big and disruptive explosion in your life and leave the debris radioactive for a long time).

To give you a clearer idea of what I am talking about, I next describe spiritual aspects of several projects that have had a profound influence on me.

Spiritualist healers in Mexico

In the late 1960s I was a young psychologist teaching in the Department of Anthropology at the University of California at Riverside, and I agreed to teach an ethnographic field methods course. I had never done ethnographic

fieldwork, never had a course in it, and felt like a fake. Fortunately, a colleague, Michael Kearney, who was carrying out an ethnographic exploratory
study of spiritualist healers in Baja California, Mexico, invited me to join
him. I thought that in working with him I would learn what he did as a
researcher and also begin to develop my own sense of how to do field
research. I had two years of high-school Spanish, but he was fluent. Riverside
is a few hours commute from the area just north of Ensenada where Kearney
was exploring, and so we could drive down there on weekends or between
school terms.

I went to Baja California as a quantitative behaviorist, a counter of
behaviors, a tester of hypotheses, and determinedly uninterested in studying
or perhaps even talking about anything that could not be operationally
defined and measured through reliable and valid measurement techniques.
I soon became acquainted with about a dozen healers, all of them older than
me and all but one of them a woman. I attended religious, healing, and other
kinds of services where a spiritualist and members of the congregation made
contact with spirits. I observed a number of efforts at healing — for example,
healing of chronic physical pain, alcoholism, grief, a sprained toe, or a child's
eating of dirt. As a positivist researcher, I was soon generating hypotheses
and counting things — for example, how the spiritualists' *templos* (typically
two room dwellings with an altar and seating for perhaps 20 people) seemed
to be places for newcomers to Baja to find community, or how some people
who seemed to be healed in the spiritualist *templos* only became in desperate
need of healing after arriving there.

At the same time, the *templos* were giving me a new sense of how people
experienced faith and their spiritual connections with others. There was a
strong sense of good will and camaraderie among the 10 to 30 people who
attended the typical healing/prayer event. People might be strangers to one
another, and their interactions might be formal and limited, but there was a
sense of support and of being a fellow human being in the same boat riding
rough seas. As one sign of the camaraderie, during healing sessions each
person present brushed her or his hands together in a gesture of brushing
good will, healing, and hope toward the person being healed. When someone
was possessed by a bad or even uncommunicative spirit, others were caring
and disturbed. I know that some who were in attendance were skeptics, but
still there was a sense that faith in God was fundamental.

I was in my late 20s then and trying to learn how to be a good teacher
of ethnographic field methods. I was also hoping that I would grow as an
ethnographic researcher and collect enough data to publish something original and insightful. Underneath my professional interest in what was going
on, I also was there as a Jew who had no interest in being converted to
someone else's religion. But what I experienced there that I hadn't anticipated was a feeling of something like envy for the small face-to-face *templo*
gatherings where people cared about one another, welcomed strangers, and
worked together to heal. I also think the faith of some congregants and the

yearning to reach faith of the others was moving to me, though I didn't connect it to myself.

I fled the spiritualist healing scene prematurely, because I had been "told by Christ," speaking through one of the spiritualists, to start healing. I knew my Spanish wasn't up to the task, but even more, I felt that it would be unethical for me to try to heal because I didn't have the same beliefs as the spiritualists and the members of their congregations. Who I now am would have handled the situation very differently. I would have confessed my lack of faith and (what everyone knew) my incompetence at Spanish and asked people to allow me to stay in their world as I had been, and to try, in good faith, to come to fit in better. But I left, feeling ashamed of abandoning Mica, the healer with whom I had been most in contact, feeling unwilling (or maybe not even competent) to learn Spanish well enough, and also feeling uneasy about taking knowledge away from good and caring people without giving back as much as I might have. One thing they gave me was a new openness to spirituality, to situations in which spiritual things happen, and to understanding the realities of people whose beliefs are different from mine.

Balinese Hinduism

In the late 1970s I became interested in the culture and family life of the people of the island of Bali in Indonesia. As a person who was still heavily involved in quantitative research and who had little experience with qualitative research, I wanted to engage in ethnographic fieldwork that would enable me to become an established and successful ethnographer. From what I could tell from reading about Bali, I thought I could learn a lot from the Balinese about emotional expression and emotional control in close relationships. I was eligible for a sabbatical year away from my regular academic duties, and I thought that spending it studying Balinese social life and emotionality could be very interesting. In preparation for possible field work in Bali, I spent a year learning basic *bahasa Indonesia* (the national language of Indonesia, though not the language of home life of many Balinese at that time). I read everything I could find about Bali that was written in English, and tried to learn from things written in Dutch and *bahasa Indonesia*. I obsessed about how to make the most of my time in Bali. I was still a loyal positivist, so I had no interest in going to Bali to learn without presuppositions. On the contrary, I theorized constantly about Bali, based on what I had read and deduced, and tried to figure out how to test the hypotheses that followed from that theorizing. I then scheduled a four-week summer trip to scout out the possibilities of spending a year doing research in Bali. I arrived in Bali with a long list of situations which I, as a hypothesis-testing qualitative positivist, wanted to observe, and I also carried a long list of questions I wanted to ask Balinese people.

Everything about Bali fascinated me. At the same time, being in Bali was the first time in my life I was on my own in a place where I had to get along in a second language. That challenged me to know myself better and to be aware in ways that were new to my thinking about the meanings of my connections with people. Also of great importance to my spirituality, Balinese Hindus (the vast majority of the people I got to know during my weeks in Bali) live a life saturated with spirituality. There was not a segregation in time, place, activity, or situation between what was spiritual and what was not. Spiritual activities and thoughts were part of everyday life and everyday understanding. As I moved toward seeing the world through Balinese eyes, I could feel the ways in which a smile, a causal interaction, an artistic action, or a random event could be understood as saturated with spirituality. There was a way that my experience in Bali made it easy to think/feel that everything I do has a spiritual element.

My short stay in Bali was incredibly rich for me and gave promise that a year of research would be well worth the effort. But it was also too hard a place for my family to be for a year, and I didn't want to be separated from my family that long. So I pulled back from my interest in Bali, but at the same time my experiences there had changed me.

I returned from Bali feeling that it was important for me to become more knowledgeable and involved in the beliefs and practices of my own religion, Judaism. Starting out in nearly perfect ignorance, I became involved in the campus Jewish congregation. It was a place that welcomed ignorant people, and I learned a lot there. Not only had my experiences in Mexico and Bali opened me to what Judaism as represented in the campus congregation had to offer, they freed me to feel, imagine, and think in open, spiritual ways while reading religious texts and singing religious songs. It is now easy for me to feel Jewish religious practice in a way that gives me what I think is a spiritual high. What I feel includes a sense of the people around me, history, and on my good days something closer to the whole universe than I could feel before my experiences in Mexico and Bali. I doubt that I could have reached that place without my experiences in Mexico and Bali.

I also need to say that I can appreciate, enjoy, and find affirming religious practices in other religions. For me, the connection of a congregation, the spirituality in religious music, the meanings of prayer, and the power of religious symbols can come from all sorts of religious bases. Maybe I could have reached that place without my experiences in Mexico and Bali, but I doubt it.

Bitter, bitter tears

Because I decided not to spend my sabbatical year in Bali, my next qualitative research project was a sabbatical year study of grief and family life, based on my readings of published and unpublished 19th-century diaries. It led to a book: *Bitter, Bitter Tears*.[18] In preparation for the project, I had read or skimmed a handful of unpublished diaries in local archives and hundreds of published diaries. While on my sabbatical year off from regular academic duties, I looked at hundreds of unpublished diaries at historical archives in the U.S. and Canada. I also read or skimmed hundreds of additional published diaries. Looking for material on grief and on family life, I wanted to get a daily texture of life and feelings, capture the writers' most private thoughts and feelings, and to get those things from a time before Americans and Canadians were influenced by the ideas of modern psychology. I was still a positivist, counting this and that, defining variables, and measuring them. But I was also aware that many diaries gripped me for more reasons than that they were an interesting data source.

When reading a diary, I could feel that I had come to know the diarist. The diarist could become understandable, real, and even predictable in the way that a close family member is predictable. That kind of knowing of another person is a connection that I might have felt as spiritual even if the diarist were alive. But knowing that the person had long been dead, getting to know the person better and becoming something of an expert on that person, augmented the spiritual connection for me. This is one of those things that is tough for me to put into words, but there was something about coming to know, and sometimes to like and care, about somebody who had been dead for many years that made a kind of bridge of feeling, empathy, understanding, and knowledge across the unbridgeable gap between the living and dead. At times I felt that it was a religious calling, of human contact and affirmation, of caring and even nurturing (though how can I nurture someone long dead?) to read every word of a diary.

One reason I became interested in unpublished diaries was that I had methodological concerns about published diaries. In particular, I believed that all published diaries were edited, and feared that often what was edited out was the kind of personal and emotional material that interested me most. Also, at the time I did the research, published diaries were disproportionately of famous men. I wanted to know about the lives of ordinary people (partly because I feared that famous people might keep diaries knowing that there was a good chance that what they wrote would someday become public knowledge), and I wanted to know more about the lives of women.

Some archives supply diary readers with a record of every person who had ever looked at each unpublished diary. I'm sure the archivists gave readers those records so each reader could learn who else was interested in, and perhaps writing about, a particular diarist. For me, those records were part of why reading unpublished diaries had a big spiritual impact on me. At times I found myself reading a diary of a person to whom I could feel drawn, one whose diary was full of life and feelings, and I could tell from the archival record that no one had looked at the diary in years. I might be the only living person who knew the diarist. That seemed like an enormous responsibility and privilege, to be the connection between living humans with a long dead diarist.

On my first visits to the archives I traveled alone. While there, I would typically scan dozens of manuscripts, hunting for diaries rich in family and loss material. I would arrive at an archive each day at opening time and stay until closing time. I would return to my motel room exhausted, but also profoundly affected by the ways I had spent the day in contact with death and life. Everyone whose diary I had looked at that day was dead, and yet every diary was filled with life and with everyday concerns. I have vivid memories since early childhood of stewing about death in various ways. So it's not as though my visits to the archives were facing me with new issues about personal mortality, but still, by the end of those day-long encounters with death I found myself wondering and feeling a lot about what was meaningful in my life, if anything, and what was not. Those were dark times for me, depressing and physically and emotionally exhausting, but I'm glad I asked the existential questions I did.

I was making good progress toward my research goals and good progress on the existential quest that reading the diaries had put me on. But I decided, after my first visits to archives, that I needed more connection with the living if I was going to come through the diary research without sliding down into a deep well of depression. I valued the spiritual quest but needed to take better care of myself against the emotional risks of the quest. So on each of my next two trips to an archive I brought along one of my two sons (ages 14 and 10). I also took one of them because I loved them a lot and was missing them. Finding ways to entertain a child in an historical archive might be worth a chapter for a book on humor in research. Trips to historical archives are not good ways for children to spend their time. So I decided, after traveling with each of my sons, to limit my subsequent choice of archives to ones near where I had friends, relatives, or scholarly colleagues. That meant that I could spend part of my time with live humans.

I weathered the year of wandering among archives well enough, and I can say, two decades later, that a number of the diarists are still with me. I think about them. I remember them when certain things come up. I have been thinking about writing novels that take off from several of the diaries. I still think now and then of editing one unpublished diary (of a woman whose entries run from girlhood into late middle age and whose accounts of her personal struggles moved me) for publication.

In a deeper sense, I think my relationships with people in my life — not only my family and close friends but everyone I meet — have been affected by my sense, profoundly reinforced by the diaries, of how ephemeral life is. I want to make the moments count. Perhaps that is something I would have come to simply by living life, but I think the diary readings was a crash course about life and death that would not easily be duplicated in the normal middle-class American life.

Grieving families

Within a few years after carrying out the diary study I had made qualitative, in-depth interviewing my research method of choice. I think my experiences in Mexico and in Bali and reading all those 19th-century diaries pushed me in that direction, partly by adding to my skepticism about the value of positivist research. I had come to think the quantitative, positivist research with which I was familiar typically produced generalizations based on an excessively superficial understanding of people. However, it took a number of years for my research approach to evolve to what it is now: research based on interviews that involve deep, phenomenological exploration with people about their closest relationships.

The study of farm families who had lost a family member in a fatal farm accident was the first interview study in which I was a primary interviewer.[21,25,26] Up to that point I had been captured by an industrial production model of research projects in which the investigator is a manager and graduate assistants gather the data, but I pushed myself, and had been pushed by people who knew me well, to become an interviewer. With the farm accidents study I became a primary interviewer. Becoming an interviewer enabled me to use all that I knew and felt to maximize my understanding of people. It meant that my research writing would be fully informed by that understanding in ways that went beyond what I learned from interviews carried out by others. This was so because I had come to realize that despite carefully thought out, predetermined schedules of interview questions and despite interviewer training, every interviewer carried out her or his own unique interviews. So it was no longer legitimate to think of myself as an effective manager. If I wanted interviews that spoke to what I most wanted (even if I didn't know at the start of a study or a particular interview what I most wanted), I had to do the interviews myself. I also thought I would be a better manager if I was an interviewer, because I would learn as I interviewed about what it was I most wanted — for example, what questions should be asked, what topics I wanted to dig at, what unanticipated ethical issues I wanted us to be more sensitive about. This is not to say that the graduate students who have interviewed for me haven't done excellent work. They have. But I thought it necessary for me to do interviews to better understand the interviews others did, to improve the focus of later interviews in a study, and to write as astutely as possible.

I also wanted to be a primary interviewer because I wanted to meet people face to face, to see them in their home setting, to carry out the informal interactions that occur prior to and after the formal interview and in the breaks that may punctuate an interview, and to have the experience of being fully present and fully open as the interview went on.

So I interviewed grieving families. The interviews were intensely moving to me — the stories people told, their pain and tears, the intense couple and family interactions. Interviewing in depth on issues related to a death, I would come home emotionally spent but also full to overflowing with thoughts and feelings. There would be a lot that I wanted to talk about and also a lot that I needed to feel/think by myself. Fortunately, my wife, Sara Wright, is understanding and is a great listener. So I could talk with her at length about what I had seen and heard and what I was experiencing. That doesn't mean that I could put everything into words, or that I had in the first hours and days after an interview come to the places the experience would eventually get me to spiritually. But I needed to talk because of all that was boiling inside of me and because I needed to get the clarification and crystallization that comes with putting experiences and feelings into words.[17] I also needed to let myself feel (and often to cry) about the people who had talked with me, the person whose death they focused on, and the frailty of human life and of our closest connections. At times I needed to feel and think in solitude, but the interactions with Sara were really important. (And so were the interactions with Terri Karis, the graduate student who was also interviewing on the farm accidents project.)

The spiritual effects of having done those interviews were enhanced by going back to visit some of the respondents more than once and by having later phone conversations with some of them. I had escaped from the restrictive idea that the only way to do research was to do exactly the same things with every person studied. In escaping, I had been freed to better know the people I interviewed, to have them know me better, to see them in various situations, and to learn how our previous interaction(s) had affected them. Everybody I interviewed in that study is still inside me. I think about them often, and they are part of my frame of reference for dealing with the world. The people whom I saw more than once especially affect me that way.

When I try now to say what happened to me spiritually through that study, I again think I cannot put it all into words. I think a lot of what I gained was about me: about me as a person who has grieved, who will grieve in the future, and who will die, and about me as a person who has a responsibility (perhaps even a vocation) to listen to and understand people talking about pain and life. So many people I interviewed in this study (and in the next one I'm about to relate) said that they had never before talked with anybody in such depth about their experiences, had never put it all together, and had never been heard as they had been heard by me. That, for me, was part of the spirituality of the study — being a catalyst for people. Another part was my sense of how important it is as a human to be fully present as a listener and caring questioner. I think that it was during this study that I became clear with myself that there is and should be a spiritual dimension to the interviews. That doesn't set interviews apart from other relationships. I'd been clear for a long time that there was a spiritual dimension to every interaction with my children, my wife, my students, and many other people. But up to this point I had kept the door closed to spirituality in the research interview — I think because that had seemed to me to be unscientific, maybe too selfish and scary.

Subsequently, I was a primary interviewer on a project focusing on parents who had experienced the death of a child.[5,22,23] As with the farm accidents study, I would come home from every interview boiling with emotion, thoughts about the people who were interviewed, and what I might call spiritual obsessing. By then, the thoughts and feelings were familiar, and I'm sure for my wife the post-interview rush of words and feelings from me was also familiar. The study carried me (like a tide carries a piece of driftwood) further on my spiritual journey. I'm sure a key part is that many of the people I interviewed were themselves wrestling with important spiritual questions about the meaning of life and of having experienced enormous adversity.

Why doing qualitative research can be dangerous

So far, I have emphasized the positive in what I experienced as spirituality and spiritual growth while carrying out qualitative research. But it wasn't all positive. Qualitative research can be dangerous, precisely because of the spiritual pieces of it.

Working on your own issues through your research

In a sense, all my qualitative research involves a focus on my own issues and/or the issues of my collaborators. I think that social research can be much more insightful when the researcher has relevant experience — though the researcher must often put personal experience aside in order to be open to the realities of others. I also think it is good to research issues that one cares about greatly, because qualitative research usually requires a great deal of time and effort. Without strong personal interest in a topic, it would be hard to sustain focus and effort. But working on your own issues also means that you are more vulnerable to the spiritual impact of your research. Thus, to the extent that grief and death have been central personal issues for me, researching issues of death and grief makes it likely that almost anything that comes up in an interview could rock me spiritually. For example, I remember an interview early in the fatal farm accidents study, where I was weeping along with the couple who was talking about the aftermath of their son's death. Here's what they were saying:

> W: We haven't really talked about it. It's a, it's, the few times that we have brought it up, it's such a hurt that we just haven't gotten into it, you know.
>
> M: More or less start crying, the two of us, and then hug each other.
>
> W: Umhm.
>
> M: and just (he laughs). We just loved him that much.
>
> (From Rosenblatt, Paul C., Ethics of qualitative interviewing in grieving families, *Death Studies*, 19, 139–155, 1995. With permission.)

My tears were partly out of empathy for them, but also reflected the impact on me of their acknowledging the distance between them, their love for the child, their pain because of his death, and the ways that they could still connect although they said that they had not been able to talk about it. My attention was on them, but at the same time I knew there were deep personal places in me that were being touched. An interview like that is challenging to carry out, partly because I can't let it become an interview about me and my personal issues. Still, although I wanted to be there humanly for them in the best possible way and to do the best possible interview with them, I also knew that they were touching me in deeply spiritual ways. By then, I had had enough experience with such interviews to trust that I could get to my own spiritual reactions to the interview later — not only immediately after the interview but in the months and years afterward. But I knew that the interview was relevant to my own issues concerning life and death, parenting, emotion in marriage, overcoming marital barriers, grieving, and much else.

You change but significant others might not

Any life experience can lead one to change in ways that might produce greater differences from significant others. I don't want to say that there is something uniquely relationship-challenging about the spiritual changes that come with qualitative research, but in my experience, the ways that qualitative research rocked the boat for me spiritually and changed me ranked right up at the top as something that was a big change for me but not for significant others. I would come back from an interview, a visit in Baja, or from reading a diary, and feel needy, vulnerable, confused, anxious, or something else that was unusual for me. The spiritual impacts and the change process (as I reconstruct now what was going on then) were disquieting. My neediness and disquiet made me different from how I had been and, at times, led me to want things from significant others that were new to our relationship and that a significant other was not necessarily able to provide. Especially in the early stages of all this, there were difficulties in my closest relationships. It could be experienced as, in a sense, pushing for the close relationship contract to change. And I could become upset, not only about what was going on with me spiritually, but about what I thought I wasn't getting (understanding, empathy, companionship in going through changes…). I suppose this kind of thing is always an issue in a close relationship, because people often do not change together, and the changes of one may challenge the other and the relationship. But it was certainly an issue for me. The part that is easiest to talk about was how my renewed involvement in Judaism was in a sense a violation of the marriage contract in my first marriage and drew my children into something religious that my wife at that time hadn't bargained for. But there were other parts that were equally significant, no less a problem, and much more difficult to put into words.

Another part of what was difficult was that I had for too long allowed myself to be trapped in disciplinary worlds, models of publishing research, and ways of thinking that made the personal pieces of what was going on in my research inappropriate to write about or even to talk about. Partly because of the influence of postmodernism, feminism, and other forces I mentioned at the beginning, and partly because people such as Kathy Gilbert, the editor of this book, are inviting us to speak about these things, I can talk now. But for a long time, it was much more lonely and confusing to experience spiritual churning and change and not know if others did, or to have somebody to help me think through things. Even now, I know that talking about spiritual processes I experience in conjunction with my research makes some people treat me as though I have a disgusting contagious disease or that I had just confessed that my closest relationships are with rocks.

All accounts of journeys simplify. And particularly when people are on journeys they did not know they were on and might not have bargained for, the story of the journey must be an after-the-fact reconstruction. That is certainly what I offer in this chapter. And in the reconstruction I leave out a lot of doubts, confusion, fear, and frustration. Perhaps the worst omission was something I briefly touched on in talking about the study of 19th-century diaries. The processes of what I here label spiritual change were getting me into some very depressing places. The depression, in retrospect, wasn't all bad because it was about understanding things that were wrong with my life as I was living it and about changes I needed to make. But the depression was still very heavy, and I had no idea how long it would last or where it would take me.

The account of my journey has also not said a lot about my nightmares as I wrestled with thoughts and feelings about death, life meaning, and what to do about the parts of my life that no longer fit where and who I had come to feel I wanted to be. This account also has not said a lot about how much the changes made me feel regret, guilt, shame, and even despair about where I had been before. For example, once I had more of a sense of how crucial human community is, I felt profound regret about having chosen an occupational route that almost guaranteed that I would never live close to my family of origin and the friends I had grown up with. Or, to take another example that is relatively easy for me to say in print: when I became much more involved in Judaism I felt regret bordering on despair about having chosen an occupation that made it very difficult to be a fully observant Jew.

I also haven't said much about the ways that carrying out qualitative research can lead one into deep pain, rage, envy, and other strong feelings that are hard to live with and may hang on for a long time. For example, the diarists whose entries led me to like, admire, enjoy, and respect them had all died. Although I knew a diarist would die, I could be weeping tears of grief as I neared the end of the diary or read a yellowed obituary that was part of the archival records about the person. Similarly, I could come home from interviewing a grieving family, hurting for them. Years later, I still have pangs of grief and challenging dreams about some of those people and their losses.

Conclusion

My field research experiences in Mexico and Indonesia opened me spiritually to the origins of my own culture, in myself, and in others. The field research experiences and my experiences interviewing individuals and families about heavy things in their lives have changed me as a human in relationship to other humans, have changed how I view myself and others, how I am a faculty member, community member, and a family member, and changed who my friends are. Some of those changes have been hard and have had side effects that may appear undesirable from a lot of angles. But I like who I have become and am only sorry it took me so long to get here. I also realize that this is a journey that will continue, so where I think I am now is not where I will be in the future. I also realize that the realities I write about here could have been written about in very different ways. My narrative, like any other, is not free-standing but is linked to context (which includes my perception of the "assignment" Kathy Gilbert has given the authors of the chapters in this book), perceived audiences, the language and cultural resources available to me, the impression I want to make, my sense of what a "good" chapter would be, the perspectives I take given my gender, race, class, occupation, religion, sexual orientation, etc., and lots else. (See Holstein and Gubrium,[7] for an extremely useful account of how to think about narratives from a postmodern, constructionist perspective.)

Conceivably, I could tell very different stories about myself and tell them in very different ways. And, conceivably, there is much more that you can do with my narrative than simply try to make sense of the surface content of what I have written.

This chapter is written with a lot of self focus and may give the reader the mistaken belief that I also have been quite self-focused in my qualitative research. On the contrary, I think I only gained what I did from my qualitative research because I focused outside myself and worked hard to understand the realities of the people who provided the data. In my experience, it is primarily in getting outside of myself that I grow. And the only way to do good qualitative research is to step outside of yourself and pay attention with your eyes, ears, heart, and mind to what others have to say. Put another way, our capacity for self-transcendence, which is at the core of spiritual growth, is very closely linked to our capacity for understanding and knowing.[6]

I don't mean to imply that good qualitative interviewing requires a single-minded focus on somebody else. Quite the contrary. I think good qualitative interviewing requires one to be able to move among various internal perspectives and, at times, to be working simultaneously with several of them. One must be fully tuned in to the people one is talking with, but one must also be thinking about what to make of what is being said, what is being left out, one's emotional reactions, how others in the room are reacting, how what is being said is linked to other things the speaker said, how what is being said is related to one's hypotheses, theories, and previous interviews, how what is being said may need to be clarified through further questioning, what the next question might be, how the tape recorder is doing, how much time is left for the interview, and dozens of other things. It's part of being a good interviewer (and, I assume, a good therapist) to acquire the capacity to be working at many levels at the same time. I think it's also part of being a good interviewer to save extensive self-focus for other situations.

I may tune in briefly on my personal feelings and thoughts, and I may self-disclose to an interviewee if that seems appropriate, but personal spiritual change is something I wouldn't let myself get into during the interview. It would be unproductive, insensitive, and even bizarre to turn an interview into spiritual therapy for the interviewer. I think of qualitative research as relational, so I intend to be real and fully present with the people being interviewed, but there are limits to where I will go during an interview.

I also need to say that the topics I have chosen to study — for example, intimate family life, grief, facing the end of a way of life — are topics that would quickly get many of us to questions of ultimate meaning and value. Had I chosen topic areas that were not so closely linked to the deepest meanings in people's lives, for example, breakfast cereal preferences, it probably would have been more difficult for me to travel the spiritual journey I traveled. And yet, I believe that interviewing people about almost anything could take one on a spiritual journey. By this I mean interviewing with integrity, curiosity, observant attention, a sensitive ear, and a determination to know the interviewee well always has the potential to create a deep and spiritually meaningful interaction.

Achieving a spiritually meaningful interaction can be more complicated when there is a power differential in the relationship, and the older "scientific" model of interviewing was one in which the interviewer had the power in all sorts of ways. In that model, the interviewer is the creator of the situation and the observer of it; the interviewer gives next to nothing of herself or himself but demands a lot from the interviewee. From a social constructionist perspective, everyone in an interview situation is at work co-constructing one another. This is especially so when the interview is less "scientific" and more a matter of mutual give and take, with the interviewee having a lot of say about where the interview goes. From that perspective, perhaps the spiritual growth I experienced was because of the ways the people I interviewed collaborated with me in constructing my identity, values, and sense of self. It is difficult to generalize, but I remember specific interviews in which I knew people were defining me in ways that surprised me. For example, I vividly remember an interview in which a couple defined me as somebody of substantial wisdom, insight, depth of understanding, and even healing power. I might have denied that I had such special qualities, but I didn't make a big point of my denials, and it is possible that my sense of growing spiritually is entangled in the "shoulds," reframing, and expectations bound up in the identity others were co-constructing with me.

For anyone who is thinking there is some sort of testable hypothesis in this chapter about the effects of qualitative research on the researcher, I should point out that at the same time I was doing these various projects I was experiencing a lot else. People important to me died, my body grew older, I got to know many people I hadn't known before, I read hundreds of books, and I co-parented three children. Also, during the course of my spiritual changes my sense of what knowledge is and my sense of what the search for knowledge is like changed to one in which objective truth was impossible. For me, knowledge comes out of the interaction between the knower and what is known. That change in me as "scientist" made it necessary for me to look at what was going on with me. Or maybe the changes in how I think about knowledge and knowing came out of the changes that I experienced in my qualitative research. At any rate, in this chapter I tell a story that emphasizes the spiritual effects on me of qualitative research, and I think it is a true story, but it is probably not possible to separate the effects of the research from everything else.

I should also point out that, in some civilizations, spiritual journeys have been a theme in accounts of life experiences for thousands of years. I am among other social researchers who have seen their interactions with the people they have studied as at the heart of a spiritual journey.[15,28] Like all metaphors,[10,20] the journey metaphor highlights certain things — for example, progress along some path, the possibility of there being a destination, the interaction of the person who is journeying with the environment through which the person travels. But like all metaphors, the journey metaphor obscures things that do not fit; perhaps the ways that one is on many different journeys simultaneously, the ways that wherever one gets to is inherent in one from the beginning, the circularity in what is described as a journey, the ways that there are no essential connections among the places that seem to be connected in an account of a journey, and the ways that life is not a journey but a biological development process.

I am aware that my account of my spiritual journey is much more about relationships with people than about relationships with God, and yet I know that for many readers a spiritual journey should by all rights be about God. To them, I would say that I do not experience a distinct boundary between people and God. In fact, when some people talk about their encounters with God there is a sense that God is everywhere, including inside people.[27] So, for the reader who expected to read more about God in this chapter, my recommendation is to read the chapter again, seeing God in the things I have said about my encounters with others and even about my struggles with myself.

A positivist, a "scientific" researcher, might take this narrative as, at best, a set of hints about what could be studied systematically. My narrative suggests that certain kinds of experiences cause certain kinds of changes in a researcher. For example, I would never have done the projects I have mentioned in this chapter in order to achieve spiritual changes. The changes were unanticipated adjuncts to my work at trying to acquire and publish useful information about human social life. Maybe that's often the way it is with spiritual change. It comes when one is fully focused on making a difference in the world. That's one of a number of hypotheses this essay suggests. A positivist could test the hypotheses suggested in this essay, carrying out longitudinal research with a representative sample of qualitative researchers. In the end, there might be a theory about spiritual change. That theory could then be pushed to further tests, could serve as a warning or an invitation to prospective researchers, and could feed into a more comprehensive theory about spirituality and spiritual change.

From another perspective, my account of the ways I think I have changed could be taken by somebody who values "objective," "scientific," positivist research as evidence that qualitative research cannot be trusted, in part because the researcher keeps changing. However, I think of my changes in an entirely different way. I think my changes have made me a more sensitive research instrument in that I have become a better listener, one who is more open to the realities of the people I interview, and one who can do a better job of representing to readers the diversity, complexity, contradictions, and ambiguity of what people have to say. At another level, we all are changing all the time — variations in mood, health, life experiences, and a million other things. No matter what kind of research one does, those variations have the potential to influence one's research activities — everything from initial planning to a final summing up of whatever one thinks is worth reporting.

Since my own orientation has become so postmodern, social constructionist, feminist, and qualitative, I would prefer that a reader see in this chapter suggestions about possibilities, not certainties or "truth." For example, don't assume you won't be affected spiritually, and don't assume if you are affected spiritually that it won't have an effect in lots of other areas of your life. From my perspective, I think that, in some sense, suggestions about possibilities are all we can ever get from any social research — from the narrative of a single case to the most carefully done, quantitative, "scientific" study. So from my perspective, there is no point in trying to use this chapter to test the ideas in it scientifically. The ideas are already here. Take them as suggestions about possibilities, about things that might happen to you and to others. Take those suggestions as being about spirituality in qualitative researchers or about spirituality in anyone. But please also take the comments that qualify and frame what is said in the chapter. Suggestions about possibilities always come from some sociocultural location and always are constrained by language choices and much else. We achieve "knowledge" in this business, but I think we have reached a point where the "knowledge" is always provisional and inseparable from the linguistic, social, and other contexts in which it arises. And the "knowledge" is not only about what is said but also about how it is said.[7]

References

1. Collins, P.H., *Black Feminist Thought: Knowledge, Consciousness, and the Politics of Empowerment*, Routledge, New York, 1991.
2. DeVault, M.L., *Liberating Methods: Feminism and Social Research*, Temple University Press, Philadelphia, 1999.
3. Gergen, K.J., The concept of progress in psychological theory, in *Recent Trends in Theoretical Psychology*, Baker, W.J., et al., Eds., Springer-Verlag, New York, 1988.

4. Gubrium, J.F. and Holstein, J.A., *What is Family?*, Mayfield, Mountain View, CA, 1990.

5. Hagemeister, A.K. and Rosenblatt, P.C., Grief and the sexual relationship of couples who have experienced a child's death, *Death Stud.*, 21, 231–252, 1997.

6. Helminiak, D.A., *Religion and the Human Sciences: An Approach Via Spirituality*, State University of New York Press, Albany, 1998.

7. Holstein, J.A. and Gubrium, J.F., *The Self We Live By: Narrative Identity in a Postmodern World*, Oxford University Press, New York, 2000.

8. Ibanez, T. and Iniguez, L., Eds., *Critical Social Psychology*, Sage, Thousand Oaks, CA, 1997.

9. Kvale, S., *InterViews*, Sage, Thousand Oaks, CA, 1996.

10. Lakoff, G. and Johnson, M., *Metaphors We Live By*, University of Chicago Press, Chicago, 1996.

11. Lather, P.A., *Getting Smart: Feminist Research and Pedagogy with/in the Postmodern*, Routledge, New York, 1991.

12. Miller, G. and Holstein, J.A., Eds., *Constructionist Controversies: Issues in Social Problems Theory*, Aldine de Gruyter, New York, 1993.

13. Mishler, E.G., *Research Interviewing: Context and Narrative*, Harvard University Press, Cambridge, MA, 1986.

14. Parker, I., Ed., *Social Constructionism, Discourse, and Realism*, Sage, Thousand Oaks, CA, 1998.

15. Quinney, R., *Journey to a Far Place*, Temple University Press, Philadelphia, 1991.

16. Rosenau, P.M., *Postmodernism and the Social Sciences*, Princeton University Press, Princeton, NJ, 1992.

17. Rosenblatt, P.C., Ethnographic case studies, in *Scientific Inquiry and the Social Sciences*, Brewer, M. and Collins, B., Eds., Jossey-Bass, San Francisco, 1981, 194–225.

18. Rosenblatt, P.C., *Bitter, Bitter Tears: Nineteenth Century Diarists and Twentieth Century Grief Theories*, University of Minnesota Press, Minneapolis, 1983.

19. Rosenblatt, P.C., *Farming Is in Our Blood: Farm Families in Economic Crisis*, Iowa State University Press, Ames, 1990.

20. Rosenblatt, P.C., *Metaphors of Family Systems Theory: Toward New Constructions*, Guilford, New York, 1994.

21. Rosenblatt, P.C., Ethics of qualitative interviewing in grieving families, *Death Stud.*, 19, 139–155, 1995.

22. Rosenblatt, P.C., *Parent Grief: Narratives of Loss and Relationship*, Brunner/Mazel, Philadelphia, 2000.

23. Rosenblatt, P.C., *Help Your Marriage Survive the Death of a Child*, Temple University Press, Philadelphia, in press.

24. Rosenblatt, P.C., Interviewing at the border of fact and fiction, in *The Handbook of Interviewing*, Gubrium, J. and Holstein, J., Eds., Sage, Thousand Oaks, CA, in press.

25. Rosenblatt, P.C. and Karis, T.A., Economics and family bereavement following a fatal farm accident, *J. Rural Community Psychol.*, 12(2), 37–51, 1993.

26. Rosenblatt, P.C. and Karis, T.A., Family distancing following a fatal farm accident, *Omega*, 28, 183–200, 1993–1994.

27. Rosenblatt, P.C., Meyer, C.J., and Karis, T.A., Internal interactions with God, *Imagination, Cognit. Personality*, 11, 85–97, 1991.

28. Schlegel, S.A., *Wisdom from a Rainforest: The Spiritual Journey of an Anthropologist*, University of Georgia Press, Athens, 1998.

chapter seven

Emotions as analytic tools: qualitative research, feelings, and psychotherapeutic insight

Jennifer Harris and Annie Huntington

Contents

Introduction

All aspects of life have emotional components, including the academic production process that involves us as researchers in actively seeking to understand aspects of human behavior and/or the world around us. However, mainstream, arguably *malestream*, approaches to research theorizing and practice have often ignored or marginalized the importance of emotions in the research process.* This is as a result of technical, rational, and instrumental responses dominating practice. This approach has been challenged, both explicitly and implicitly, by feminists, postmodernists, and interpretivist

---

* As Emma Wincup states in Chapter 2, denial of the feeling element of research work, and the emotional labor needed to manage this, are common practice in mainstream accounts of the research production process.

social scientists who have questioned the validity of key concepts (for example, the vision of the researcher as a dispassionate, objective observer), thereby providing us with alternative frameworks for research practice.[29,31] These have opened a space within which a focus on the emotional labor involved in producing research is legitimate.[1] However, taking emotions seriously into account requires an ongoing commitment to articulating alternative narratives about the nature of academic production. To be clear, doing so involves critical reflection, as a focus on emotions in research is often seen as a specialist concern or of marginal interest.[31] Work focused on emotions continues to transgress normative expectations in academia.[29]

When we are involved in researching people's social lives we are likely to experience a range of emotional responses as they (re)present their world to us. As much social research has been conducted with what can be characterized as marginalized or oppressed groups[27] who experience various sorts of disadvantage, it is unsurprising that researchers come to vividly co-experience participants' everyday realities. In this way their cause may becomes our cause. Their anger at discrimination may become our anger. In many ways, the isolation we feel as a result of our difference from them may be mirrored by their isolation from the mainstream society of which (to varying degrees) we are a part. However, we may also struggle with our emotional responses if those we study articulate experiences and responses that we find difficult or even impossible to identify with.[6] The thorny problem of being different from research participants may result in intense isolation experiences or feelings of relief. These emotional responses, we claim, can be used in either positive or negative ways. If used positively, this can become an exercise in acknowledging the differences between the self and the other, which may ultimately shape the emerging text.* However, used negatively, the result may be a spiral of individual despair, uncritical analysis or an inability to complete our projects. As Bordo[3] states, "The pleasure and power of 'difference' is hard won; it does not freely bloom, insistently nudging it's way through the cracks of dominant forms."

As writers focus on the involvement of the researcher as an active participant in the production of research,[33] or articulate new and innovative ways to present the products of research,[36] consideration of the importance of the researchers' emotional worlds is brought into sharp focus. If researchers are active participants then they are very likely to have emotional responses. However, acknowledgment of this dimension of academic work often remains invisible. This is possibly because any focus on emotions traditionally was, and still largely is, directed at the research participants'

---

* This idea, we freely acknowledge, has parallels with one element of "emotional intelligence." Goleman[20] describes how "marshalling emotions in the service of a goal is essential for paying attention, for self-motivation and mastery, and for creativity." Clearly, Goleman does not stretch the idea as far as we do below, inasmuch as we claim here that the emotional "memory" of fieldwork can assist in the production of rich, accurately empathic texts.

experiences of involvement in research projects. If we view academic work as a production process, then a lack of focus on the emotional labor involved in the production of knowledge could be seen as an example of the dominance of masculinized vision.[25] As feminists have demonstrated,[46] emotional labor is often ignored or devalued across a range of arenas, as it is seen as women's work. In addition, as Pyke and Agnew[41] state, the complexity of the debate is often arbitrarily simplified. Deeply entrenched and embedded ideas about the emotionally labile nature of women (the "irrational subject"), serve to ghet-toize any discussion of emotions as inevitably suspect if one is interested in producing "real" science. What Jaggar[29] terms the "dumb" view, (positivist and neopositivist approaches), continues to legitimize the myth that we can somehow access the uncontaminated truth if we edit ourselves out of the research process. This means academic exploration of emotions in social research remains difficult. However, if we take emotions and emotional labor seriously into account, then we open a space within which we can explore practical strategies to work with our emotional responses. In addition, we bring to light aspects of experience that may be particularly problematic for novice researchers or those engaged with substantive topics that are likely to engender strong reactions (for example, sexual abuse of children).

Within this chapter we explore ways in which we can usefully attend to the emotional aspects of research practice. This analysis will include refer-ence to theoretical insight and understanding from the therapeutic arena, as it applies to research practice. In summary, we argue that the process of acknowledging the emotional impact of events in our practice enables us to analyze reflexively the differences between the values and experiences of the self and the other. This emotional work, though sometimes painful, enables us to evaluate our practice in far greater depth — tapping into levels of meaning that are not always apparent when we negate the importance of our emotional responses. Similarly, emotions can act as "markers" — a way of reliving events when analyzing our practice or responding to research subjects.* Our contention then is that emotions not only shape interactions between ourselves and research participants, but they also can be utilized to recollect and relive the detail of significant events. As Okely states, "Inter-pretations are attained not only through a combination of anthropological knowledge and textual scrutiny, but also through the memory of the field experience, unwritten yet inscribed in the fieldworker's being."[39]

However, this focus on the importance of emotions in research practice does not lead us to romanticize this aspect of experience as the only or primary filter to define and understand the world around us. Solipsism and egoism need to be avoided, as we attempt to articulate and explore alterna-tive ways to access and report on the social world.[42]

---

* This potential is also discussed by Emma Wincup in Chapter 2. In her account she acknowl-edges the potential to use her emotional responses constructively throughout the research production process.

Social research and emotions

In general, those writing about qualitative research practice recognize that data collection, analysis, and subsequent theorizing are highly dependent on the researcher's personal positioning. As Dale et al. states, "...in qualitative research that involves ethnography or in-depth interviewing the researcher becomes the research instrument, with all the results being filtered through her perception and understanding of the social situation in which she is working."[8]

Research involves the collection, interpretation, and representation of information. This work is influenced by the person of the researcher in many overt and subtle ways. This includes work within research arenas where findings have traditionally been presented as if they are hard scientific facts.[43] If we acknowledge that the person of the researcher is important within the production process, then we have to ask ourselves questions about the ways in which our "person" is influential in the process of producing research narratives. In addition, we need to ask questions about the impact of external forces, as these shape our emotional responses and affect our performance as researchers. The hegemony of research traditions, institutional constraints, and the impact of political agendas affect the ways in which we experience and mobilize our emotional responses as we attempt to explore aspects of the social world through research practice. For example, emotions can be used as institutional resources, to achieve instrumental ends when deployed by social actors in specific situations.[38]

Once we focus on the impact of the self, we face core concerns for theorists debating many contemporary social issues: the relationship of structure to agency, and the constraints on autonomy.[18,40] Structure here is taken to equate with rules and resources,[32] while agency links to the extent to which individuals can make choices unconstrained by overt or covert processes, either within social institutions of modern society or in terms of their intrapsychic worlds and interpersonal relationships. As Giddens states, "Human beings produce society but they do so as historically located actors, and not under conditions of their choosing."[18] This assertion is echoed by Plummer, "We human beings are social world makers though we do not make our social worlds in conditions of our choosing."[40]

If we consider issues of structure and agency, whether we focus on philosophical, psychological, or political explanations of our identities,[19] we need to acknowledge that academic work is not just an esoteric activity. Research is a potential instrument of emancipation or domination.[4,18] Knowledge is ignored, suppressed, or promoted to meet particular agendas[14] within the inevitable hierarchies that exist in unequal societies.[40] Accepting that all knowledge is contingent, partial, and open means we need to explore the extent to which our personal commitment to specific ideological or political agendas influences our judgments about the legitimacy of specific accounts of phenomena. As Everitt and Hardiker state, "The structures and processes through which apparently objective facts and subjective experiences are generated and filtered need to be interrogated."[13]

Again, these concerns are as valid when discussing the process of academic production as they are when focusing on its products. Jaggar[29] states that epistemological justification of a hierarchy, wherein reason dominates emotion, defines the space within which knowledge can legitimately be constructed. This serves to silence those "who are defined as culturally the bearers of emotion and so are perceived as more subjective, biased and irrational."[29]

When exploring the nature of research production, we need to question dominant visions and take emotions into account at the ontological, epistemological, theoretical, and practical levels. Production of reliable knowledge does not rest on the exclusion of the emotional dimension of human existence. In fact, to the contrary, inclusion of a focus on emotions can enrich the process of producing knowledge. However, it is only possible to assert the importance of emotional labor if we challenge the dominance of the Western philosophical tradition wherein emotions are judged to be an anathema to academic production.[29] This is not as easy as may be presumed. Focusing on emotions in social research, especially our own, is problematic in practice,[11] as a concern with emotional issues can result in a Catch-22 situation for women situated as academics. If we focus on emotions, we fit a stigmatized stereotype and may end up positioned below rational cognitive actors in the hierarchy constructed on the basis of dominant conceptions. The irony here is that to fight to be taken seriously is to provide justification for one's own subjugation, but to deny the importance of emotions is to be co-opted into the masculinized mainstream, aligning oneself with those with power as a way to stay safe. The struggle to "break out"[44,45] means we have to challenge assumptions. To do this we need access to personal and professional resources. In particular, we have to find ways to challenge any internalized oppression we may experience as we struggle to manage potentially stigmatized aspects of our person(al) identities.[21] If we take just one example, that of lesbians, we can see that the internalization of negative societal messages about homosexuality can fundamentally impact how any woman positioned as a lesbian experiences her sexuality. In the research arena, a focus on emotions is often difficult, as we have to deal with any internalized repression that says this aspect of experience is not a legitimate focus for academics.

If we acknowledge the importance of the emotional dimension, we might then work to an alternative hypothesis. We could assert that women's ability to work with emotions is a strength, not a problem to be overcome.[3] The feminization of emotional labor means that women have a head start — they are more used to working with the contradictions and tensions of emotional experience. However, in doing this, we need to avoid situating women within privileged epistemological territory.[24,35] Rather, we must assert the importance of a particular form of reflexive practice, focusing upon our emotional and cognitive responses when in the role of researcher. This is the operationalization of what has been termed the "feminine gaze"[25,29] and more stereotypical masculinized concerns are key in understanding this converse construct. Through this means, finding ways to consciously work with our emotional responses is a version of "good" practice. However, this is a far more difficult and messy process where repression is real, (arguably in all social institutions within contemporary society across cultures), than can be captured in the sanitized versions of research presented in many academic texts. Yet to omit this level of experience leaves us without language of expression or cultural markers when we engage in research activity. Perhaps the contradictory and ambiguous feelings likely to be engendered when we take emotions seriously, that link to our experiences of being powerful and powerless,[3] are a block to integrating a focus on emotions as analytic tools in routine ways.

Exploring emotions in qualitative research

At the practical level, emotion, especially recollected traumatic emotion, can be a useful analytic tool when undertaking field work.[23] However, conceptions of emotion and feelings are contested at the definitional level. As Jaggar[29] states, defining emotions is difficult as the term is used to describe diverse phenomenon in ordinary usage. Identification and discussion of emotion is a culturally and historically specific project. In addition, O'Brien[38] argues that emotions are usually analyzed as individual artifacts or expressions of systemic relations, rarely both. As well as these definitional difficulties, attempts to delineate the differences between emotions and feelings are evident. However, consensus is lacking as to how emotions and feelings differ. The subject of emotions may be treated as an abstract concept, which although it acknowledges that feelings are intrinsically linked, still manages to speak of emotion without addressing the impact of the former upon the researcher and fellow participants. For example, within an exploration of "feminist fractured foundationalist epistemology positions," Stanley and Wise[44] claim that, "Emotions, the product of the mind, can be separated, at least at the level of theoretical discussion, from feelings, rooted in the responses of the body; cold and pain are feelings, love and envy are emotions."

This position reflects Jaggar's stance in that she argues that emotion and feeling are often conflated as synonymous.[29] For Jaggar, "feelings" often link to physiological sensations while "emotions," although associated with physiological responses, include more conscious aspects. This mean they are not easily divorced from other human faculties like cognition.

Stanley and Wise[44] acknowledge that separation for theoretical purposes is "by no means simple," but why should we wish to attempt to understand either emotions or feelings in isolation? These issues, of course, depend entirely upon particular definitions of "emotions" or "feelings." They seem to imply that emotions are somehow more "pure" than feelings; that perhaps feelings could be considered also to be experienced by animals, whereas emotions could not. Their definition then, implies that while emotions might be considered theoretical "meat," feelings are comprised of raw animalistic urges which are not amenable to study. How could we understand an emotional response in isolation from the feelings that generated it, and further, why should we wish to do so? We would argue that both feelings and emotions are data and therefore should both be considered "meat."

Setting aside for one moment whether we can and should operationalize this stance (and how it is done), there is the more pertinent issue of why feeling and emotion are discussed in ways which do not acknowledge how they affect the researcher and the process of research in general. Therefore, while some feminists have acknowledged the importance of emotion and the sin of its omission from Cartesian ontology,[44] the more practical issues of how emotion and feeling affect the research process have not been addressed and the issue of their usefulness has been largely ignored. This is rather curious, since emotion, in some feminist texts, is linked to very fundamental issues such as ontology and epistemology[44] and indeed Jaggar suggests "...that emotions may be helpful and even necessary rather than inimical to the construction of knowledge."[29]

The importance of emotion is acknowledged within some feminist texts but it would seem that the discussion of its effects, impacts, and possible coping strategies are missing from the debate. Why is this so? It would seem that one reason is that the debate concerning emotion takes place at a level largely divorced from the practicalities of empirical research and therefore does not engage with the applicability of derived theory or its usefulness for such purposes. Alison Jaggar, for example, puts forward an eloquent case for the importance of emotion in feminist epistemology,[29] but concentrates almost exclusively upon the ways in which emotion is defined by authors without considering what the effects of using one definition over another might be within empirical research projects. The only hint that this is an interactive process is contained within this quotation from Jaggar:

> Just as observation directs, shapes and partially defines emotion, so too emotion directs, shapes and even partially defines observation. Observation is not simply a passive process of absorbing impressions or recording stimuli; instead it is an activity of selection and interpretation. What is selected and how it is interpreted are influenced by emotional attitudes.[29]

There are similar curiously curtailed debates in the literature from the sociology of the emotions. Arlie Russell Hochschild[25] structured her debate of feelings and emotion around the idea of "images of actors" which, she claimed, were divided into conscious/cognitive and unconscious/emotional. However, she did not appear to admit of the possibility of the sentient actor which she proposed. This now rather dated work obviously still has implications today for the ways in which emotion is discussed, or is invisible within the literature, since the focus is exclusively on the "researched." This is a second criticism of existing literature on emotions: the feelings and emotions discussed are entirely those of the researched group and no mention is made of the researcher's emotions at all, let alone whether there might be some usefulness in the latter as data. Notable exceptions to the limited discussion of the practical implications of a focus on emotions and feelings are available.[11,12] However these are problematic in that these accounts can be said to unhelpfully sit within the romantic philosophical tradition.* The focus is on the explication of personal experience in ways that are problematic at two levels. First, it obscures the extent to which emotions may be political and social constructs.[38] Second, the approach advocated is likely to be emotionally demanding and time-consuming in ways that raise questions about the boundary between research and therapy. Whether we believe approaches like interactive interviewing[12] are morally defensible, practically feasible, or sound research practice is open to question even if we accept the importance of focusing on emotions in social research. However, a focus on emotions as a key source of insight about social life may clearly lead us to experiment with imaginative ways to undertake research that may produce different sorts of knowledge about the world around us.[1]

Overall, the most detailed discussion of the effects of emotional encounters in the field is to be found in the more broadly based fieldwork literature. The best of these for our purposes is by Kleinman and Copp,[31] who take the issue of the researcher's emotions as their central topic. In fact, though, there are very few places in which these issues are discussed in general and even fewer where authors are prepared to be honest about such issues as feeling angry toward participants.[31] However, the authors' major contribution to the field undoubtedly is that they not only give emotions and feelings pride of place in their writing, but they also acknowledge the ways in which the researcher's feelings and emotions can be used for the purposes of data analysis:

> As I look back, my anger served as an inequality detector. This detector however, is fallible; we should use it to test whether or not we are witnessing an injustice. But we can only test this hypothesis if we first acknowledge

---

* We would like to clarify that our understanding of this term focuses on the anti-enlightenment approach to experience that stressed the importance of the nonrational or even irrational aspects of human behavior. Although feelings, imagination, and experience are important, they are not the only filter to understanding the world. In addition, personal experience needs to be understood with reference to the inevitably political aspects of living if we are to do more than focus on the individual alone.

> such feelings as anger. Facing my worst fear, that I was
> unempathic, led me to articulate my analytic position
> and explain why it fit the data better than some other
> perspective.[31]

Engaging with the emotional effects of ethnographic field work is traumatic and reliving experiences through the notes can be equally distressing. It is also still comparatively rare that authors include descriptions and analyses of intense emotions experienced in field work. However, doing so is crucial if we are truly to call ourselves participants in settings.[22] Acknowledging the emotional effects of such field work encounters enables us to relive our time in the field in a vivid manner.

This tapping into the experiences at the level of emotions is a powerful tool for the analysis and subsequent reanalysis of fieldnotes. Our argument is that researching in this way enhances understanding of everyday social relations of groups. In addition, actively reflecting on our emotional responses while in the field allows us to systematically record our responses as they happen. These can be a source of rich analytic insight later, as previously discussed. They can also provide the readers of the products of research with valuable indicators about the process of conducting any piece of work. For example, Huntington's[28] fieldnotes from a research project (focused on children and families social work practitioners' responses to changing employment conditions and work practices) describe the processes in detail:

> I feel swamped and inadequate as a researcher. I don't
> feel I am keeping up with what is happening and find
> it difficult dealing with the conflictual accounts I am
> being presented with. I get confused as I am sympa-
> thetic to individual interviewees but when I try to
> analyze what I think is going on and relate it to a wider
> picture I just feel lost. When I do feel like I am getting
> a grip on what is happening it quickly slips away. This
> feels hopeless. Am I mirroring some of the atmosphere
> in the children and families division, as people at all
> levels struggle to make sense of what are often im-
> posed changes in ways that allow them to actually
> continue to work in the authority and possibly in social
> work. Alternatively, am I reading things in the way I
> am because that is how I feel about social work?

Such active reflection can provide us with a valuable source of information and an analytic tool that is not confined to work in the field. We can engage in a similar process when interviewing people, whether individually or in focus groups. This allows us to monitor our responses, as we develop the ability to trust yet remain skeptical of our reactions, as a route to generating better-quality data for analysis. Making use of the researcher's emotions and feelings as data, and the important ways in which doing so can enrich research reports and writings, would move the debates within feminist thinking and the sociology of the emotions out from the level of theory and into the realm of practice. Such a move could only be mutually beneficial. It is clear that the benefits of this approach to research practice are being actively explored by sociologists undertaking empirical research in a number of areas.[37]

Working with emotions: lessons from psychotherapeutic practice

If we accept that managing, containing, and working with emotions is a legitimate part of the research process, then we need to focus on practical strategies to operationalize this intention. Exploration of bodies of literature, situated within differing professional domains, provides us with a rich source of information with which to explore the ways in which we might manage our emotions and use them as analytic tools. We do not have to reinvent the wheel. However, to mine other theoretical seams requires that we have some basic understanding of what might be useful to us as there are many approaches to therapeutic practice. Major schisms are as evident in approaches to theory and practice in the therapeutic arena as they are in the world of research. As Clarkson[5] highlights, there are three major psychotherapeutic traditions: psychoanalytic, behavioral, and humanistic/existential. These traditions have differing philosophical antecedents, theoretical explanations for emotional and psychological experiences, practical approaches to therapeutic work, and differing goals in practice. In addition, all three traditions are internally inconsistent, as theorists have modified original concepts and approaches to practice. For example, Freudian theory has been redefined in ways that lead to major schisms within the psychoanalytic community: "There is nothing in psychoanalysis, from its most general principles to its most specific observations, that can be said to be universally accepted."[15]

Ultimately defining and working with therapeutic theory for practice is as political as working with theory for research practice. In both arenas some approaches are presented as if they are core, or mainstream, while others are seen as marginal or peripheral.[42,47] Although clearly identified traditions are evident in either the research arena[35] or the therapeutic arena,[5] not all approaches are afforded the same status or legitimacy. This is not just linked to the applicability or usefulness of approaches but also reflects more primal concerns that link to how we treat the "other" — that is, those who we cannot see reflected in ourselves. As Clarkson[5] identifies when discussing differing approaches to therapeutic work, "It is also true that issues of power, ideology, money, status, employability and snobbery play a significant part in such territorial axis."

This concern was echoed by May[35] when he discussed contemporary definitions of the self. These are often framed in oppositional terms and this practice can easily lead to oppressive acts at a time when "striving for identity through difference is now a common act."

Acknowledgment of the differing schools or therapeutic traditions is important here as it allows us to focus on the exploration and application of psychotherapeutic knowledge and understanding without claiming that this is an easy or simple process. Schisms and tensions exist, and this insight hopefully sensitizes us to the importance of keeping an open mind — rather than resorting to deterministic, reductionist, or exclusionary claims about the application of psychotherapeutic knowledge for research practice. More specifically, we will draw on work from within the humanistic tradition/existential tradition as this is familiar to one of us* and more generally reflects our stance to the production of knowledge (as interpretivist, feminist applied social scientists). Psychodramatic understanding of the nature of truth rests on acknowledgment that we co-construct narratives — "...psychodramatic truth: this truth does not aim at any precise demonstration of facts, but at a subjective representation of reality."[34] We believe this is of particular relevance here.

Alternatively stated, there is no presumption that a definitive version of reality is available to the therapist or client. Rather, knowledge is contingent, partial and open — generated in the here-and-now of our interactions that are shaped by the there-and-then experiences within our formative and subsequent social atoms. (This term refers to the matrix of internalized relationships — real/fantasized, past/present — that link to any individual's life experience/s).

For us, there is no such thing as the neutral, objective, dispassionate, but interested observer who can access the truth — either as the expert therapist or the scientific researcher. Our reference to this therapeutic tradition for the purposes of this chapter, however, does not imply that this approach is more valid or applicable than others. It is an expression of the art of the possible.[10] This means that it provides a vehicle to discuss issues of interest if we want to move beyond articulating the rationale for attending to emotions and the benefits of doing this. It provides us with tools to focus on how this might be done rather than why it might be necessary to do it. Those with other training or experience might then actively reflect on the skills and knowledge that they might transfer if they occupy the role of social researcher and want to take emotions seriously into account in their practice.

Overall, a number of key concepts might be useful as we work with emotions as researchers:

1. Acknowledgment that even if we engage in interpersonal interaction on the basis of hierarchical role relationships (for example, therapist and client or researcher and research partner), we hold reciprocal roles as equal human beings.[48]

* Dr. Annie Huntington is a qualified psychodrama psychotherapist.

2. Cognitive exploration of emotions enriches our understanding of interpersonal processes and our own responses. We can use our emotions **and** cognition to understand the world around us.[30]

3. We can develop the ability to reflect actively on our own emotional responses, even while we are experiencing them.[26] Feelings or emotional responses do not have to detrimentally influence the exercise of our intellectual ability.

4. Tension and conflict in human relationships is inevitable. Experiencing difficult emotional responses is normal within everyday human relations. The key, as we interact with others, is to own rather than justify our feelings.[2,30]

5. Understanding something of our own emotional landscape, including our past experiences and the impact these may have for us in the here and now, is not self-indulgent. Rather, it allows us to understand and yet be free of the past. Knowing how we come to be as we are facilitates our active management of emotional responses.

Utilizing emotions within the research process: general issues

'Within a research context there are a number of ways in which we might apply these lessons. At the most pragmatic level, we might usefully reflect on the ways we can facilitate interviewees' narratives to ensure both they and we get the most out of the time spent together. We might reflect on the ways in which we can make the rhetoric of partnership and participation in the research process real; in that, we ensure that we treat people as active subjects not passive objects. For example, creating space for interviewees to discuss what is important for them, taking their concerns seriously into account, attending to nonverbal as well as verbal cues in interactions or ensuring that respondents understand the boundaries of the interaction, may all facilitate the collection of high-quality information as they have an impact on the emotional tenor of the interaction. Additionally, attending to the emotional component of interviews when engaged in the interaction and when transcribing and analyzing material may highlight points of particular impact in the narratives. For example, signs of emotions may include overt expression (the loss of emotional control, crying), problems answering questions (repeated requests for restatement of questions), self-reporting of emotional impact ("this is really difficult as it still hurts"), and problems narrating their account (fragmented and seemingly random discussion of events). These indicators may in turn alert us to significant aspects of accounts to which we need to pay particular attention.

At another level, attention to our own emotional responses within interactions, including the work we do with the products of our face-to-face contact with research participants, provides us with further opportunities to reflect on and work with emotions as analytic tools. For example, within interviews we may actively reflect on the feelings generated for us and attempt to work and learn through them. In this way we may access differing levels of understanding through the formulation of further questions that link to our understanding of our emotional as well as cognitive responses. This may lead us to gather more comprehensive, or different, sorts of data as well as increase our sense of achievement and personal well-being. This could have particular salience within certain sensitive or difficult fields that inevitably exact an emotional toll upon us (for example, research into self-harm or loss and death). Furthermore, attention to our own emotional responses outside interviews may enable us to continue with difficult or emotionally draining research work. This could become part of a structured self-preservation program, whereby simple strategies such as taking a walk or a bath act as balm to aid the conservation of our mental health. Our emotions can become sounding boards[23] — a means of creatively analyzing and evaluating our work from a position of grounded safety.

Ultimately, our goal should move from the justificatory position in relation to utilizing emotions in the production of research, which to date has dogged the field, toward discussion and debate concerning the best means and mediums for portrayal of such data. Indeed, we have little to lose and everything to gain from such a move.

Applying understanding — an example

Whether we are involved in interpersonal interactions for therapeutic or research purposes, the interview is a core site for interactions. Further, as Stuhlmiller (see Chapter 4) argues, the principles and techniques used within interviews are similar. Thinking through and working with the emotional rather than just the technical aspects of our interactions in interviews is an important, and we would argue often neglected, consideration. This is most notably highlighted if we consider the nature of interviews around substantive topics that generate particular ethical, technical, **and** emotion management concerns — for example, with sexual offenders. In these instances we need to think very carefully about the ways we structure the interactions with interviewees to get their story. More specifically, we need to ensure that we have the intrapsychic, interpersonal, and organizational resources available to maximize the chances that we produce good-quality information while keeping both the interviewee and ourselves "safe."

Again, if we look to the therapeutic literature, we can access some interesting learning points of interest to the social researcher. For example, thinking about people's motivations and rationales for engagement in therapeutic interviews highlights the complex and at times contradictory concerns that may be evident. This is clearly an issue when working with sexual offenders — whose overt and covert agendas for engagement need careful exploration.[9] Similarly, a research interviewee's motivation for engagement may be complex and/or require some teasing out. Not everyone is interested in passing along wisdom and knowledge that may be useful to others, and if their agenda is questionable we may have a moral and ethical duty to ensure that such information is not disseminated. Active reflection within interviews with interviewees about motivational issues may offer important insights that frame the narrative generated and speak to the substantive concern that is being explored. However, to focus on such concerns we need to legitimate our interest in process as well as product within research interviews. This leads us to another area of interest.

Exploring issues around interviews with someone who has sexually offended alerts us to the need for extra care in terms of the monitoring and management of our emotional responses even as we engage in the interview experience. If we do not engage in such work, we are in danger of being embroiled in unhelpful interactions that will affect the quality for the information we generate — for example, if we are "sucked into" colluding with the offenders behavior in some way this will affect the nature of the information we collect. Powerful processes are inevitably generated when we allow ourselves to really open up to people who have committed violent sexual acts and these have to be acknowledged and managed, not ignored or denied, as we work to describe, explore, and explain such behavior and/or societal responses to those who engage in it. Alternatively stated, the narrative that is heard is constructed in the interactions between the participants, and attending to emotional concerns even as we do the work is likely to yield different, arguably better-quality information if we take our emotional responses seriously into account.

Concluding comments

Whether we decide to actively work with the emotional aspects of research practice is a theoretical and personal decision that links to our beliefs about epistemological, methodological, and pragmatic questions about how we come to know the social world. To focus on the affective components of research requires that we restore the emotional dimension to conceptions of rationality and challenge the assertion that objectivity and unemotionality are synonymous. As Cook and Fonow[7] assert, "This aspect of epistemology involves not only acknowledgment of the affective dimension of research, but also recognition that emotions serve as a source of insight or a signal of rupture in social reality."

Operationalizing this stance suggests that we must resist attempts to consign discussion of emotions to the "feminine" sphere as it demeans their importance within the production of good research. Emotions affect all human beings at all stages of the life course and there is every reason to suppose that they are as important within the production processes described here as they are in everyday life. Understanding and analyzing emotional responses, therefore, facilitates access to deep meanings, hidden agendas, and ultimately, "thick description."[16] The effort of working analytically with emotions is justified both in terms of the quality of work we produce and of the importance of maintaining our own emotional well-being.

References

1. Bendelow, G. and Williams, J., Introduction: Emotions in social life. Mapping the sociological terrain, in *Emotions in Social Life: Critical Themes and Contemporary Issues,* Bendelow, G. and Williams, J., Eds., Routledge, London, 1998, XV–XXX.
2. Blatner, A., *Acting-in: Practical Applications of Psychodramatic Methods,* Springer, New York, 1988.
3. Bordo, S., Feminism, Foucault and the politics of the body, in *Up Against Foucault: Explorations of Some Tensions between Foucault and Feminism,* Ramazanoglu, C., Ed., Routledge, London, 1993, 190–210.
4. Bulmer, M., *The Uses of Social Research: Social Investigation in Public Policy Making,* George Allen and Unwin Ltd., London, 1982.
5. Clarkson, P., The nature and range of psychotherapy, in *The Handbook of Psychotherapy,* Clarkson, P. and Pokorny, M., Eds., Routledge, London, 1994, 3–27.
6. Condor, S., Sex role and "traditional" women: Feminist and intergroup perspectives, in *Feminist Social Psychology,* Wilsonson, S., Ed., Open University Press, Philadelphia, PA, 1986, 97–108.
7. Cook, J. and Fonow, M., Knowledge and women's interests: Issues of epistemology and methodology in feminist research, in *Research Methods: Exemplary Readings in the Social Sciences,* Neilson, J., Ed., Sage, London, 1990, 1–21.
8. Dale, A., Arber, S., and Procter, M., *Doing Secondary Analysis,* Unwin Hyman, London, 1988.
9. Deacon, L. and Gocke, B., *Understanding Perpetrators, Protecting Children: A Practitioners Guide to Working Effectively with Child Sexual Abusers,* Whiting Birch, London, 2000.
10. De Vaus, D., *Surveys in Social Research,* 3rd ed., UCL Press, London, 1991.
11. Ellis, C., Emotional sociology, *Studies in Symbolic Interaction,* 12, 123–145, 1991.
12. Ellis, C., Kiesinger, C., and Tillman-Healy, L., Interactive interviewing: Talking about emotional experience, in *Reflexivity and Voice,* Hertz, R., Ed., Sage, London, 1997, 119–149.
13. Everitt, A. and Hardiker, P., *Evaluating for Good Practice,* BASW/Macmillan, London, 1996.
14. Foucault, M., *The Archaeology of Knowledge,* Routledge, London, 1972.
15. Frosh, S., *The Politics of Psychoanalysis,* Macmillan, London, 1987.
16. Geertz, C., *The Interpretation of Cultures,* Fontana Press, London, 1993.
17. Giddens, A., *New Rules for Sociological Method: A Positive Critique of Interpretative Sociologies,* 2nd ed., Polity Press, Oxford, U.K., 1993.
18. Giddens, A., Living in a post-traditional society, in *Reflexive Modernization: Politics, Tradition and Aesthetics in the Modern Social Order,* Beck, U. et al., Ed., Polity Press, Cambridge, 1994, 56–106.
19. Glover, J., I. *The Philosophy and Psychology of Personal Identity,* Penguin, London, 1989.
20. Goleman, D., *Emotional Intelligence: Why It Can Matter More Than IQ,* Bloomsbury, London, 1996.
21. Goffman, I., *The Presentation of Self in Everyday Life,* Penguin, London, 1959.
22. Hammersley, M. and Atkinson, P., *Ethnography: Principles in Practice,* Routledge, London, 1983.

23. Harris J., Surviving ethnography: Coping with violence, anger and frustration, *The Qualitative Report*, 3(1), [On-line]. Available: http://www.nova.edu/ssss/QR/QR3-1/harris.html.

24. Harris, J. and Paylor, I., The politics of difference, in *The Politics of Research and Evaluation in Social Work*, Broad, B., Ed., Venture Press, London, 1998, 31–43.

25. Hochschild, A., The sociology of feeling and emotions: Selected possibilities, in *Another Voice: Feminist Perspectives on Social Life and Social Science*, Millman, M. and Kanter, R. M., Eds., Anchor Books, Garden City, NJ, 1975, 280–307.

26. Holmes, P., *The Inner World Outside*, Routledge, London, 1992.

27. Hornsby-Smith, M., Gaining access, in *Researching Social Life*, Gilbert, N., Ed., Sage, London, 1993, 52–67.

28. Huntington, A., Differing Perceptions of Legislative and Policy Change in Children and Family Services: A Vertical Analysis, unpublished Ph.D. dissertation, University of Central Lancashire, Preston, U.K.

29. Jaggar, A., Love and knowledge: Emotion in feminist epistemology, *Inquiry*, 32, 151–175, 1989.

30. Kellerman, P., *Focus on Psychodrama: The Therapeutic Aspects of Psychodrama*, Routledge, London, 1992.

31. Kleinman, S. and Copp, M., *Emotions and Fieldwork*, Sage, London, 1993.

32. Lash, S., Reflexivity and its doubles: Structures, aesthetics, community, in *Reflexive Modernization: Politics, Tradition and Aesthetics in the Modern Social Order*, Beck, U. et al., Eds., Polity Press, Cambridge, 1994, 110–175.

33. Lee, R., *Doing Research on Sensitive Topics*, Sage, London, 1993.

34. Marineau, R., *Jacob Levy Moreno 1889–1974. Father of Psychodrama, Sociometry and Group Psychotherapy*, Tavistock/Routledge, London, 1989.

35. May, T., *Situating Social Theory*, Open University Press, Buckingham, U.K., 1996.

36. Mienczakowski, J., An ethnographic act: The construction of emotion in consensual theatre. Paper for the 4th Int. Social Science Methodology Conf., University of Essex, Essex, U.K., 1996.

37. Newton, T., The sociogenesis of emotion: A historical sociology?, in *Emotions in Social Life: Critical Themes and Contemporary Issues*, Bendelow, W. and Williams, S., Eds., Routledge, London, 1998, 60–80.

38. O'Brien, M., The managed heart revisited: Health and social control, *Sociolog. Rev.*, 42, 393–413, 1994.

39. Okely, J., Thinking through fieldwork, in *Analyzing Qualitative Data*, Bryman, A. and Burgess, R., Eds., Routledge, London, 1994, 18–34.

40. Plummer, K., *Telling Sexual Stories*, Routledge, London, 1995.

41. Pyke, S. and Agnew, N., *The Science Game: An Introduction to Research in the Social Sciences*, Prentice-Hall, Englewood Cliffs, NJ, 1991.

42. Reinharz, S., *Feminist Methods in Social Research*, Oxford University Press, Oxford, U.K., 1992.

43. Shmidt, M., Grout: Alternative kinds of knowledge and why they are ignored, *Public Adm. Rev.*, 53, 525–530, 1993.

44. Stanley, L. and Wise, S., *Breaking Out: Feminist Consciousness and Feminist Research*, 1st ed., Routledge and Kegan Paul, London, 1983.

45. Stanley, L. and Wise, S., *Breaking Out Again: Feminist Ontology and Epistemology*, 2nd ed., Routledge and Kegan Paul, London, 1993.

46. Walby, S., *Patriarchy at Work*, Polity Press, Cambridge, 1986.

47. Watson, R., The illusion of psychodrama: A study in the psychology of psychodramatic representation, *British J. Psychodrama Sociodrama,* 8, 5–47, 1993.

48. Williams, A., *Forbidden Agendas: Strategic Action in Groups,* Routledge, London, 1991.

49. Williams, A., *The Passionate Technique: Strategic Psychodrama with Individuals, Families and Groups,* Tavistock/Routledge, London, 1989.

chapter eight

Collateral damage? Indirect exposure of staff members to the emotions of qualitative research

Kathleen R. Gilbert

Contents

Introduction

This chapter addresses the emotional impact of qualitative research on a group that has received little formal attention — research staff members. I

0-8493-2075-5/01/$0.00+$.50
© 2001 by CRC Press LLC

use the term "staff" rather loosely to refer to any individuals we involve in our research projects in an assistive capacity. These may be paid staff or students who are compensated in some nonfinancial way (for example, they receive academic credit or do so as part of their course requirement), as well as those who volunteer to help with a project.

Over the past several years, interest in the effects of research on both the researcher and the researched has grown dramatically. For the most part, we qualitative researchers no longer see ourselves as objective observers, chronicling the stories of others. We do not simply observe a phenomenon and report on it. Rather, we recognize that what we observe and experience as reality is filtered through various lenses, with the emotional context being one of these filters. Through our involvement and interpretation, we become a part of the process and part of the story we report.[6,13,18] This requires the *active* construction of meaning[7,20] and an *active* involvement, at several levels, in the researched phenomenon.[18] Thus, we do not document, coolly and objectively, our findings. We bring our thoughts and our emotions with us when we, as researchers, explore some aspect of social life about which we are concerned.

At the same time, we also attempt to maintain a sense of "otherness," as we construct and present what we hope is a representative image of whatever we are exploring. This sometimes results in internal conflict and heightened emotion and these emotions are often complicated by the intensity and unpredictability of the research process. This conflict is documented in the personal accounts of researchers (this can be seen in Rothman's[23] account of her study of women who had undergone amniocentesis, as well as in those of the various authors in books edited by Sollie and Leslie,[24] and Ellis and Flaherty,[5] and in many of the chapters found in this volume).

The situation is most complicated when we explore emotionally charged topics.[11,23] Ellis has argued that, when investigating emotional experiences, we should do so emotionally.[4] This enriches our understanding of the data and of our interpretation of it. In the volumes edited by Sollie and Leslie[24] and Ellis and Flaherty,[5] the contributing authors discussed sometimes surprising ways in which they connected with their work and the participants in their studies — and the ways in which this connection informed their understanding of the focus of their research. I, too, have been much affected by my work on parental loss and grief at the death of a child. I now know, at more than an intellectual level, that children can die before their parents, and that the parents may be powerless to affect the outcome of events. I recognize that time is precious and that each encounter might be our last. To another person, this statement might seem melodramatic; to me, it is now a statement of fact.

Staff members and their emotions

If qualitative research is such an emotional challenge, why do we do it? This is a question I have posed over the years to a number of qualitative researchers.

Ultimately, the responses are that, while challenging, it is also invigorating and that the benefits outweigh the drawbacks. But what about our research staff, a group that likely will include individuals who meet all of the criteria to put them at risk? Why do they become involved in qualitative research? Students may want to impress a professor or they may want to learn more about a particular topic or about the profession. Hall and Zvonkovic[15] have noted that faculty see participating in the research process as a valuable learning experience for students and encourage them to become involved. There may be other, more pragmatic reasons for researchers to use assistants. The research design may require a staff, something that is becoming more common with increased funding of qualitative projects. Faculty may respond to the pressure to "publish or perish" and, in order to reduce our workload and make more efficient use of time and other resources, engage the services of individuals to assist them in what they see as more "functional" tasks of the process.[14]

The question can be raised, then, about the genuine risk to persons hired to deal with these functional research tasks. In this chapter, I argue that characteristics of research assistants may make them particularly vulnerable to the emotions generated by qualitative research.[15] Staff members bring unique characteristics to their work on a project, which include academic status, age, life cycle stage, gender, the connection between their life experiences and the focus of the research, and their vulnerability to the emotions being evoked by the study.[14,15] They may have little, if any, research experience; they are relatively unknowledgeable about the process of doing research as well as the content area being researched; they have limited control over the way in which their role in the process is expected to be played out. All of this may be further complicated by such factors as the research design and methods used, the particular tasks they are asked to do, the makeup of the research team and their relationship with other team members, and content area of the study.

In my role of research project director, I have become aware of the impact of the life cycle stage of my assistants, particularly when they are traditionally aged undergraduates (18–22 years of age). A majority of my students (and, as a result, most of those who assist with my research) fall into this age group. Many are dealing with late adolescence issues of individuality and are searching for their professional role in life. Sometimes this works well as they develop as scholars and as individuals. In some cases, students have made career choices based on what they have learned from working on research projects. I have also seen students develop empathy and great sensitivity to others through their work with me. At the same time, I worry about the impact of the research itself on some of the students, and have had occasion to move students from one area of the project to a less emotionally challenging one. It likely will be necessary, one day, to remove an assistant from a project based on their emotional response to some aspect of the project.

These emotional effects have been observed among both qualitative and quantitative research staff, although they are expected to be more intense with qualitative methods.[11] Kitson and colleagues[17] have used structured and semistructured interviews in their studies of widows and divorcees. They noted that interviewers who were in similar life circumstances as the interviewees or the children of interviewees, were the most affected by the interview experience. In their study of women whose husbands had died violently, Kitson's group found that if the death seemed to be the end of a troubled life (a history of psychiatric problems), the interviewers personalized the death less and were able to distance themselves emotionally from it and the study participants. On the other hand, if the husband's murder occurred without apparent provocation, the interviewers found the interviews extremely difficult to handle. How could this innocent person be killed, without somehow bringing it on himself? If it could happen to him, it could happen to me or someone I care about.

This response is consistent with the "just world" theory[19] that is, at essence, a belief that the world is a fair place and that there is a reason for everything that happens. In this view, good people are rewarded and bad are punished. Using this belief, we protect ourselves from emotional overload (and from a belief in a chaotic world) by constructing explanations that attribute blame (or responsibility) to others for their misfortunes. If we are unable to find ways in which the victimized other "deserved" his fate, it is more difficult for us to find ways in which we can ourselves avoid that fate.[16]

Hall and Zvonkovic reported on their study[15] of seven female researchers who had participated in a research project in which they conducted data collection, transcription and analysis of a qualitative study of how married couples make work/family decisions. They found the following researchers were most affected: (a) researchers who were most intensely involved over a short period of time, especially if inexperienced in doing research; (b) researchers who did not have a congruence between ideology and life experience in marriage; (c) researchers with a heightened awareness of gender, oppression, and power in marriage; and (d) researchers whose emotions were evoked by the research, particularly those who did not have many outlets for venting their emotions.

It appears, then, that the degree of immersion in the research phenomenon, the ability to personalize (or depersonalize) the stories to which they are exposed, the extent of their training and preparation, and opportunities for venting are the factors with the greatest impact.

My personal story — growing awareness

Over the years, I have worked with student and nonstudent research assistants in a variety of capacities. I have supervised dissertation projects, taught qualitative methods, hired paid staff, given academic credit for research assistance, and worked with unpaid volunteer (usually student) assistants. The projects I have directed have covered a range of topics, with most of my work, as I indicated above, focusing on loss and grief.

Early in my academic career, I involved a number of young, female undergraduate students in my research. Because they had very limited background in research methods, I assigned them to what I perceived as the simplest of tasks. I was converting from one computer system to another and set them to retyping and reformatting transcripts of interviews from my dissertation project: an interview study of couples who had lost babies in pregnancy or infancy.[9] I was, at the time, working with a colleague on a book that was to come from my dissertation data[12] and wanted to have "clean" transcript copies with which to work. Not realizing the impact reading these transcripts might have on these women, I did minimal emotional preparation with them, telling them that what they read might be upsetting and if it was, they should let me know. I did not take into account such things as the possibility that students might feel a need to "keep it to themselves" because they wanted to impress me with their ability to act the role of researcher (objective, unemotional, distanced from the phenomenon[15]). Nor did I take into account their vivid imaginations and the degree to which they were able to personalize the accounts they were reading. Rather, I placed far more emphasis on technical specifications, quality and quantity of output, and time lines I established for them.

I was surprised and upset when, after their initial enthusiasm, almost every one of these students "became lackadaisical," making errors in their typing and missing deadlines. My response was more one of frustration than of concern, assuming they had become bored with the process. I tried encouragement, rewards, ultimatums. Finally, I investigated — and spoke with Ruthie,* one of the student volunteers with whom I had a good working relationship, and whose work had just begun to slide.

Ruthie's initial response was to apologize and try to explain her behavior away by attributing it to a hectic schedule. Probing further, I learned that, for her, simply reading the interview transcripts had generated strong emotions which she saw as negative and inappropriate. She described her responses with phrases like "really upsetting," "made me angry," "felt like crying," and "wanted to hit someone." Ruthie felt she had come to know these bereaved parents whose personal stories of loss she was reading and typing and she was shocked by the intensity of their grief and by their horror stories of mistreatment by others. At the same time, she felt intense frustration at not being able to do anything to help them and a degree of guilt for having thought the loss of a baby in pregnancy or as a neonate "wouldn't be that bad." It was these contradictory emotions that got in the way of her being able to work on the transcripts.

After talking to Ruthie, I sought out other current and past research assistants and asked them about their experiences. A few indicated that they had not "connected" with the research process and had, indeed, become bored. Most, however, told me essentially the same thing as Ruthie had.

---

* All names of research assistants in the article are pseudonyms.

They described dealing with strong emotions which generated aversive reactions to the stories they were reading and typing. One woman who had stayed with the project said she did so by covering all but the line she was typing while trying not to put whole sentences together. Incidentally, this sometimes impaired her ability to transcribe accurately, as she would occasionally skip over lines as she moved down the page. Because she was trying not to put the lines of text together into a coherent picture, her efforts to be self-protective impeded her ability to do a thorough and accurate job.

I also learned that these young women were motivated to volunteer by more than a desire to learn about the research process. They had gotten involved in the research projects because of an interest in human behavior and, in many cases, an interest in the helping professions. Reading these transcripts, they had constructed images which sometimes included attribution of motivations to help them understand the experience of the participants.[20] They described how they found themselves thinking about the people who I had interviewed and their babies, even dreaming about them at unpredictable times. Many described an emotional reaction that, in many ways, mirrored a grief response. Thus, the images they constructed to help themselves to make sense of the lives of the participants also triggered strong emotions, and it was these strong emotions with which they struggled.

Drawbacks for research staff

Looking at the experience of these young women in the context of our training as researchers, we might frame their response as the two extremes of over- and underinvolvement in the research process. Given their lack of expertise and experience, their minimal preparation, and their stated wish to do a thorough and complete job, this "response of extremes" should not be surprising.

Overinvolvement

Overinvolvement can be experienced as an unnerving loss of a sense of separateness, especially those new to the conduct of research.[6] It has been referred to by the term "going native" (usually used as a negative term). Certainly, for my research assistants who spoke of frightening and confusing dreams in which they delivered dead babies, and for those who developed symptoms of grief, this was a painful and frightening experience. Much like a medical student who develops symptoms of every disease studied, they empathized with and personalized the stories of participants to the extent that they, the research assistants, took on the symptoms of the participants. Experienced researchers might develop ways of balancing their involvement and/or use techniques to deal with their emotions. Neophyte research assistants, however, may be distressed because they do not understand why they are responding as they are.

Underinvolvement

One possible outcome of the research staff who feel emotionally overwhelmed by the research can be the opposite — underinvolvement, or emotional exhaustion leading to "burnout."[2] In this, the researcher feels overwhelmed by the research process and withdraws physically and/or psychologically from it. This has been described by Corsino[2] as it relates to avoidance of field work. I found the same process going on for my research assistants who were exposed to these stories only through the printed page. In each case, the sense of vulnerability and powerlessness contributes to their pulling back from participation in the research effort. The student who typed only a line at a time needed to use a type of "self talk" to psych herself up to do even this.

Research assistants also may become emotionally numb in order to cope with emotional overload.[11] This serves a protective function; however, it may also mean that stories need to become more dramatic before they generate a response. It would be wise for everyone involved in the research effort to remain vigilant for this desensitization to the "normal" yet vital details of each story. In particular, if multiple interviewers are used and if there is any latitude allowed in the interview schedule, caution must be maintained.

An alternative interpretation of the process

Another way of looking at this is to see research assistants as experiencing a secondary victimization, much like that which has been seen among therapists and direct service providers who work with traumatized individuals.[2,8,20] Also referred to as compassion fatigue[8] or vicarious victimization,[25] this indirect traumatization is identified in cases where an individual, someone providing services or a close family member or friend who was not directly affected by a trauma, exhibits symptoms of the person who was directly affected. Images and memories taken from the traumatized individual may be internalized by the secondarily affected person, such that they become a part of that person. These may be invasive, "flashing" into one's mind at odd times. They may also disrupt his/her psychological and/or interpersonal functioning.[20]

Secondary trauma is the result of one of two things: (1) the secondarily affected person's efforts to empathize with the primarily affected person, or (2) inadvertent exposure to the primarily affected person's symptoms.[10] An example would be a field worker who, after working intensely with a war veterans' group, begins to dream of her experiences of watching her closest friends die brutally in war.

I have identified three forms of secondary traumatization as they occur among spouses,[10] all of which are applicable to researchers: proximal, distal, and resonating secondary traumatization effects.

Proximal effects of exposure to the research process

This is the stress of direct contact with the primarily affected individual. Although a single, intense contact can result in proximal traumatization, so can long-term exposure, either to several individuals or through an extended time in the field. Thus, field workers and interviewers are most susceptible to this type of traumatization.

Distal effects of exposure to the research process

Distal traumatization is a vicarious victimization that results from "events in the imagination."[25] There is no direct contact with the individual who has experienced the stressor (the research participant). Instead, through such means as hearing interview tapes and reading transcripts or detailed field-notes, research assistants such as transcribers, typists, or coders are affected by their personalization of the stories to which they are exposed.

Resonating effects of exposure to the research process

In this case, two or more individuals are affected independently of each other, yet emotional displays of one person trigger emotions in the other. This might be seen in team debriefing sessions, when one interviewer becomes emotional while describing her response to the interview situation, triggering an emotional response for other team members. Research team members may feel that, rather than providing support for each other in these meetings, they add to the stress felt by each other.

My personal story — a new awareness

After my initial experience with research assistants, I was extremely concerned about the potential risk any assistant might face while working on one of my projects. In my grief projects I usually do all the interviews, and in those cases staff members do not come face-to-face with the interviewees. For a later project on grief following the termination of a wanted but compromised pregnancy, I hired a very competent young woman to transcribe the tapes of my interviews. "Jackie" would sit for hours, listening closely to tapes of women and, in some cases, their husbands or partners, who had chosen to terminate a wanted pregnancy. The pregnancies were terminated because of a lethal or serious maternal or fetal condition, and the result for many of the parents was intense, anguished grief. The interviews were emotionally challenging to do and, I was sure, would be difficult for Jackie to hear.

I decided that I would approach my relationship with Jackie in a way that was similar to the way I approach my participants. I acknowledged that she knew herself better than I ever could, so I trusted her to tell me if she felt stressed by the job and when there was anything being triggered by the interview tapes that I had not anticipated. I also told her that she might find herself empathizing with the participants to the extent that she actually mimicked grief symptoms. I also mentioned that she might experience some upsetting dreams. We agreed to touch base weekly, to debrief and to provide her with an opportunity to keep me apprized of her emotional state.

I worked with Jackie for almost a full year. She was highly dedicated and when my funding ran out, she stayed on to complete transcription of my tapes. For security reasons, we agreed that she would use my office computer for her work and she came into my office on weekends and evenings to transcribe. From my working relationship with Jackie, I confirmed some of what I had observed in my earlier assistants and learned a few new things.

As with the other assistants, Jackie did not reveal all of her feelings to me at the time she was experiencing them. The overriding fear she expressed to me was that she did not want me to become concerned enough to remove her from the project. There were no other tasks I could shift her to that did not involve working with the interview data. Because we debriefed on a regular basis and I had encouraged her to find other outlets for her emotions, she felt she was able to maintain an involvement without feeling as though she had become overinvolved and, as a result, overwhelmed. Rather than emotionally numbing herself, she immersed herself in the task and monitored herself for her response.

I made a point of putting together a warning list of aspects of each interview tape that she might find disturbing or upsetting. Although she could have read through my fieldnotes or spoken with me about the interview before transcribing, Jackie preferred the simple list.

Because Jackie also was a full-time student, she experienced heightened stress and elevated emotions around the middle and end of the semester. Exams and other course demands were the cause of this added stress and we adjusted Jackie's schedule to allow her to work as she was able. I was concerned about her transcribing for long stretches of time, but she developed a system that worked well for her. She found she could transcribe for as many as 8 to 10 hours on a Saturday, which she would then follow with time spent with either her friends or a trip home.

I found it curious that Jackie did not experience any of the adverse effects about which I had forewarned her. She did not have dreams about her own loss of a baby, as had my earlier research assistants, nor did she develop any of the symptoms of the participants in the study. Instead, she responded to the theme of the research, and to particular triggers in the interview tapes, but these were triggers not related to infant loss. I will provide two examples of what I mean. One response she had to the theme of the research was to think of her *own* birth, which had been a complicated one, in which both she and her mother could have died. She was much affected by the stories of the parents in the study, and this triggered a desire on her part to know more about her own birth. A value that she saw coming from this was that she appreciated her health while also recognizing how fragile life is. My second example is of Jackie's response to particular triggers that were not related to infant loss. In my interviews, I have found the importance of establishing a meaningful connection between the participants and myself, and I do this by describing my own losses — usually focusing on either my mother's or brother's deaths. I also acknowledge how this is different from the death of one's child, as I also know there are ways in which a child's death is unique. For Jackie, the death of an imaginary child was too abstract to consider, but the possible death of her mother or brother were difficult for her to think about. Hearing me speak of my losses over and over, and thinking about her own mother and brother, personalized the reality that their death was a possibility. For Jackie, this was one of the most stressful aspects of the project. Characteristically, she reframed her awareness in a positive way, and saw it as providing evidence that she needed to be appreciative of relationships and to be more open with her feelings toward her family members.

Consistent with what Hall and Zvonkovic[15] wrote about faculty members seeing participation in the research process to be a valuable experience for students, Jackie found this to be enormously valuable to her. During the study, Jackie's maternal grandfather, who had been ill for quite some time, died. Instead of staying at school, as would have been her decision before working on the study, she went home to be with her mother. They were able to spend time together, with Jackie's mother talking about her own childhood and her memories of her father. Jackie learned much she had never known before about her own family and developed an appreciation for the person her grandfather had been and especially as the young father she could never have known. From the stories of the participants, she found that "people need other people" and one should not be afraid to show concern.

Another way in which she benefitted from the study was that she found the direction she had been seeking in career decision-making. Since working on the project, Jackie graduated and went on to graduate school in a Master's of Social Work program with the goal of working with families who have experienced loss and trauma.

Benefits for research assistants

When exploring the literature on the topic of staff issues and emotions, I have found the emphasis to be on anticipating, monitoring for, and addressing staff problems. It was difficult to find works that addressed the benefits to our staff. Yet, as qualitative researchers we often note the value (both scholarly and personally) of taking emotional risks — being open emotionally to the lives of our participants and to their information. As can be seen in Jackie's story, these benefits can extend to our research assistants as well.

As I have mentioned before, the emotional connection qualitative researchers feel to their project and to various aspects of it, has been described as one of the most positive aspects of this style of research. Cynthia Stuhlmiller, an author in this volume (Chapter 4) who studies traumatic stress among emergency workers, has described herself to me as "really alive" when she was in the midst of a project.[27] Making a connection at both an intellectual and an affective level allows her (and others) to understand the phenomenon in a unique and deeper way. Ellis[4] has argued for such an approach when studying emotions and emotional topics. How can we possibly understand emotions from a purely intellectual perspective?

An emotional connection to the project also provides motivation for staying involved when the doldrums hit. When advising students on research, I have told them that they need to feel passionate about the project. Much as Emma Wincup (Chapter 2) was advised to find a topic that really angered her, I believe that one benefits from being intensely, emotionally engaged in a project to see it through to the end. My observation has been that, without that passion, one is more likely either to drop out of the project or become less concerned with maintaining precision.

The greatest value for research assistants is the opportunity to learn about themselves by learning about others. This has been described by my student assistants as the most beneficial aspect of the process. They felt they had benefitted from the experience, both in terms of skills and exposure to the research process. Although the emotions triggered by the process have occasionally been difficult for them, they saw themselves as benefitting from reading the stories of the participants. As project director, I found their perspective on this was informative and occasionally surprising. There were occasions, for example, when I saw someone as overinvolved, while that person saw herself as deeply involved and gaining insight on "what I should do with my life." Jackie's decision to become a social worker specializing in serving families who have experienced trauma is an example of this.

In some cases, this insight was not apparent until time had passed. Several years after assisting in a study, I received a phone call from a former research assistant who had become very angry with the bereaved fathers in my study. At the time, she found their anger and their tendency to back away from their grief to be infuriating. She could not see how these men did not simply support their wives, who she saw as the "real grievers" (a view that is consistent with the broader society, incidentally). The reason for her call was to tell me that her brother and his wife had recently lost their first child, late in the pregnancy. She wanted to "take it all back. Now I understand." She also told me that the parents whose transcripts she had read taught her what to do and both her brother and sister-in-law had told her that she was "the only one who knew what we needed."

What we can do for our staff

What can we do, as project directors, to reduce the risks to our staff while also facilitating positive effects of assisting in qualitative research? When hiring staff members, we must look for technical expertise (for example, typing speed and accuracy), experience, and enthusiasm. At the same time, I believe we also need to determine what, if any, emotional baggage they bring with them. We have an ethical obligation to prepare them for possible emotional reactions, both positive and negative, they may have to the content and/or process of the research project. I find it interesting that Brannen observed,[1] "Having sat sometimes for several hours at a time in people's homes listening to their stories and to their distress I have often thought that no psychiatrist or psychologist would work (or be allowed to work) under these conditions."

We would not necessarily want to duplicate the approaches of the therapy and emergency service fields, but their literature can be useful. A therapist, for example, who attempted to enter and become a part of the world of his/her clients likely would experience burnout quickly. Instead, Corcoran advocated the use of an empathetic stance.[2] In this, one senses the experience of the other as if it is their own, without losing the "as if" quality of the experience. His contention is that it is the loss of the "as if" quality that leads a therapist to be overwhelmed by the phenomenon, leading to burnout. A similar case can be made for researchers to retain a sense of balance in their approach to research.

Educating the staff about the phenomenon under study can reduce its impact, since the uncertainty of not knowing what to expect heightens the emotional state. O'Rear[21] described an approach to training of emergency workers that included education of the workers before they were exposed to traumatized individuals, followed by monitoring of the staff after any exposure to a trauma scene. This two-pronged approach minimizes the negative effects of exposure and can be translated to a research setting. It would be particularly valuable when training and monitoring interviewers.

A majority of the literature emphasizes emotional risk to researchers and the need to monitor for negative effects. It is important also to take advantage of the possibilities for positive emotional effects. In my orientation, I encourage prospective research assistants to maintain a reflective stance, monitoring for ways in which they learn something new about themselves and their lives from the information they encounter in the project. Research assistants also need to know that it is normal to experience stress, and that if they experience symptoms of stress it does not mean they are inept researchers.

Whenever possible, it is advisable to anticipate problems and, if possible, screen individuals out if they are clearly at risk of being emotionally overwhelmed. The list of characteristics suggested by Hall and Zvonkovic,[15] noted earlier in this chapter (academic status, age, life cycle stage, gender, the connection between their life experiences and the research focus, and their emotional vulnerability), can provide a starting point. It is important to remember that this screening for at-risk individuals is possible in extreme cases, but an ongoing monitoring of staff is essential. Everyone who works on a project has the potential for being overwhelmed.[17]

Particularly when working on emotionally challenging topics, it is important to include a segment on the emotional toll of working in this area in the orientation session. Fravel[28] informs prospective research assistants that they may find the tasks which they are assigned to be too difficult. If this proves to be the case, they know that they can ask for either a permanent or temporary reassignment.

Staff members may experience an event or events during the course of the study that changes their sensitivity to the emotional content of the study. As I learned, it may be necessary to move them to less challenging tasks.[17,26]

Interviewers need to feel emotionally supported, particularly when the topic is a difficult one. They may be concerned about their safety (legitimately) when they go out to do the interview. Kitson and colleagues[17] found interviewers to be especially affected by interviews they did with widows of men who had died violently. They found that by preparing their interviewers for each individual interview with detailed information about the interviewee, reduced the emotional response. Providing interviewers with a response segment at the end of the interview schedule may be helpful to them. Giving them one's office and home telephone number with instructions to call if they need to talk about the interview can help them to feel that they are not alone in their response. Reminding them that they might experience strong emotions, and then they might have to debrief from that, also should be included.

Coders and transcribers who work with audio or videotapes will deal with their own emotional reactions. Fravel[28] recommends encouraging them to pace themselves, restricting their work with tapes and transcripts to relatively brief periods of time.

Periodic staff meetings are an essential tool for conveying information and maintaining oversight over the project. They are particularly useful when working with a large staff.[17,22,26] Information can be efficiently shared and touching base with staff members is made easier. Kitson's group[17] took advantage of meetings to provide opportunities for interviewers to describe their emotional reactions to the interviews and to provide support for each other.

Group support meetings, possibly facilitated by a clinician,[17] can provide staff members with a structure for discussing their thoughts and feelings about the research process. They also can talk, in this structured environment, about their evolving view of the subject under study. This can serve the purpose of providing an additional "reading of the data," while it allows them to interact with each other as they make sense of their experience.[10]

It may be necessary, if research assistants feel overwhelmed, literally to walk them to help.[28] Establishing a relationship with a clinician or preplanning a response protocol will make this experience easier on you, the research director.

Students have particular issues, primarily related to the unequal relationship between students and faculty. They need to know the nature of the study and the risks they face when they volunteer to assist on a project.[15] The desire to perform the role well and to please the faculty member, concerns about grades and letters of reference, and other indirectly related factors play into their emotional response. The safeguards listed here, as well as others, should be incorporated into the teaching, training, and supervision of students.

Conclusions

In our work as qualitative researchers, we sometimes hire staff members. Just as we are affected by the emotional content of our research, so too are staff members. Unfortunately, they may be "invisibly affected," invisible to us because we fail to look or are unwilling to recognize the signs and they are unable or unwilling to let us see them. It is our responsibility, as project directors, to be aware of the possible emotional impacts, both positive and negative, and to maintain contingency plans for helping them to deal with their emotional overload. We need to integrate structures into the project that facilitate benefits and reduce risks to staff members. Finally, we need to inform them of the emotional risks and benefits of assisting in our research.

References

1. Brannen, J., Research note: The study of sensitive subjects, *Sociolog. Rev.*, 36, 552–563, 1988.
2. Corcoran, K. J., Interpersonal stress and burnout: Unraveling the role of empathy, *J. Soc. Behav. Psychiatry*, 4(1), 141–143, 1989.
3. Corsino, L., Fieldworker blues: Emotional stress and research underinvolvement in field work settings, *Soc. Sci. J.*, 24, 275–283, 1987.
4. Ellis, C., Emotional sociology, *Stud. Symbolic Interaction*, 12, 123–145, 1991.
5. Ellis, C. and Flaherty, M. G., Eds., *Investigating Subjectivity: Research on Lived Experience*, Sage, Newbury Park, CA, 1992.
6. Ely, M., Anzul, M., Friedman, T., Garner, D., and Steinmetz, A. M., *Doing Qualitative Research: Circles Within Circles*, Falmer Press, London, 1991.
7. Feixas, G., Personal construct theory and systemic therapies: Parallel or convergent trends?, *J. Marital Fam. Ther.*, 16, 1–20, 1990.
8. Figley, C. R., Ed., *Compassion Fatigue: Coping with Secondary Traumatic Stress Disorder in Those Who Treat the Traumatized*, Brunner/Mazel, New York, 1995.
9. Gilbert, K. R., Interactive Grief and Coping in the Marital Dyad Following the Fetal or Infant Death of Their Child (Doctoral dissertation, Purdue University, West Lafayette, IN, 1987) *Diss. Abstr. Int.*, 49/04A, 962, 1988.
10. Gilbert, K. R., Understanding the secondary traumatic stress of spouses, in *Burnout in Families: Secondary Traumatic Stress in Everyday Living*, Figley, C. R., Ed., St. Lucie Press, Boca Raton, FL, 1997, 47–74.
11. Gilbert, K. R. and Schmid, K., Bringing our emotions out of the closet: Acknowledging the place of emotion in qualitative research, *Qualitative Fam. Res.*, 8(2), 4–6, 1994.
12. Gilbert, K. R. and Smart, L. S., *Coping with Infant or Fetal Loss: The Couple's Healing Process*, Brunner/Mazel, New York, 1992.
13. Glesne, C. and Peshkin, A., *Becoming Qualitative Researchers: An Introduction*, Longman, White Plains, NY, 1992.

14. Gregory, D., Russell, C. K., and Phillips, L. R., Beyond textual perfection: Transcribers as vulnerable persons, *Qualitative Health Res.,* 7, 294–300, 1997.

15. Hall, L. D. and Zvonkovic, A. M., How research affects researchers: Emerging questions. Paper presented at Theory Method Preconf. Workshop, Natl. Council Family Relations, Portland, OR, Nov. 1995.

16. Janoff-Bulman, R., *Shattered Assumptions: Towards a New Psychology of Trauma,* Free Press, New York, 1992.

17. Kitson, G. C., Clark, R. D., Rushforth, N. B., Brinich, P. M., Sudak, H. S., and Zyzanski, S. J., Research on difficult family topics: Helping new and experienced researchers cope, *Fam. Relations,* 45, 183–188, 1996.

18. Kleinman, S. and Copp, M. A., *Emotions and Fieldwork,* Sage, Newbury Park, CA, 1993.

19. Lerner, M. J., Justice, guilt, and vertical perception, *J. Personality Soc. Psychol.,* 20, 127–135, 1971.

20. McCann, I. L. and Pearlman, L. A., Vicarious traumatization: A framework for understanding the psychological effects of working with victims, *J. Traumatic Stress,* 3, 131–150, 1990.

21. O'Rear, J., Post-traumatic stress disorder: When the rescuer becomes the victim. *J. Emergency Med. Serv.,* 17, 30–35, 38, 1992.

22. Punch, M., Politics and ethics in qualitative research, in *Handbook of Qualitative Research,* Denzin, N. K. and Lincoln, Y. S., Eds., Sage, Thousand Oaks, CA, 1994, 83–98.

23. Rothman, B. K., Reflections: On hard work, *Qualitative Sociol.,* 9, 48–53, 1986.

24. Sollie, D. L. and Leslie, L. A., Eds., *Gender, Families, and Close Relationships: Feminist Research Journeys,* Sage, Thousand Oaks, CA, 1994.

25. Terr, L. C., Family anxiety after traumatic events, *J. Clin. Psychiatry,* 50(Suppl.), 15–19, 1989.

26. Walker, A., You can't be a woman in your mother's house: Adult daughters and their mothers, in *Gender, Families, and Close Relationships: Feminist Research Journeys,* Sollie, D. L. and Leslie, L. A., Eds., Sage, Thousand Oaks, CA, 1994, 74–98.

27. Stuhlmiller, C., Personal communication, April, 1994.

28. Fravel, D., Personal communication, May, 1996.

chapter nine

Extreme dilemmas in performance ethnography: unleashed emotionality of performance in critical areas of suicide, abuse, and madness*

Stephen Morgan, Jim Mienczakowski, and Lynn Smith

Contents

Ethnography, health, and performance

The use of drama to influence social, political, and health practices has a long and well-understood tradition, dating from Greek and classical times.[20]

* Based upon a paper presented at the Couch Stone Symposium (Society for the Study of Symbolic Interaction: February 1999, University of Nevada, Las Vegas).

Within the modern era, we can see that the works of Dario Fo, for example, represent not only fine examples of drama, but in addition represent a significant mode of political and cultural critique. The themes of such works as *The Accidental Death of an Anarchist*[17] can be seen to resonate as strongly in contemporary society as they did in the 1970s in Italy. Similarly, ethno-drama, as first described by Mienczakowski,[25] should be interpreted as the staged performance of cultural texts developed, written, and authenticated by health care patients,* their associated health care specialists, academics, and the general public (as audience). Ethnodramatic performance allows victims of mental illness, sexual abuse or rape, substance abuse or plastic surgery, *in partnership with health care professionals and academics*, to explore and examine, through dramatic devices, what it is like from the inside looking out — that is to say, from the perspective of the patient. Simulta-neously, ethnodramas offer the potential to demonstrate and/or experience the wide-ranging emotions associated with particular pathogens and conse-quently may inadvertently unleash unanticipated emotional responses in audiences during or subsequent to a performance. Ethnodrama should be seen as a means of providing access (for self and others) to the deconstruction of knowledge acquired through firsthand (lived) experience in an endeavor to provide meaning, understanding, prevention, and perhaps catharsis and solace. Its potential lies within ethnodrama's capacity to concurrently be a viable reflexive tool for informing the provision of informed health services; a mechanism for shaping and informing political and public will; and a vehicle for emancipatory practice.[9,33] Thus ethnodramatic works can be rec-ognized as existing within a "critical" paradigm.[1]

For those unfamiliar with qualitative inquiry it would be timely to men-tion at this point that in recent times performance has been regularly used as a mode for theoretical explanation and critique in numerous disciplines outside the traditional arts arena. Evolving as part of natural experimenta-tion with and development of textual and interpretative constructions, these new performance genres are predominantly founded in ethnographic research. Often, those involved are researchers within the fields of commu-nications studies and the social sciences. Typically, their roots were located in performance and they are now devising new modes of theoretical repre-sentation related to their previous expertise and relocation into newer or different strands of social theory — Dwight Conquergood and Laurel Rich-ardson are two renowned performance text practitioners.

Reflecting the international scholarly community's response to the press-ing requirement to build research and research profiles, the new drama research culture spawned a variety of tentative research paradigms. Some aspects of these reflect approaches which are concerned with cultural critique and public explanation and remain highly enigmatic for those academics who retain a traditionalist research approach. Yet while certain sectors of

---

* Health care patients/informants are not always the exclusive focus of ethnodramas.

academia have viewed performance texts as problematic or challenging, others regard them as a means of interpreting and producing research which is publicly accessible. Contrary to previously held views that ethnographic performance was in opposition to the research values of traditional ethnographic practices, it should be noted that these newer forms of research are subjected to the same academic rigor applied to other research methodologies. Indeed, further to Denzin's comments that performed narrative must produce texts which are "accessible and performable,"[11] we believe that the essential element of accessibility to the research values as utilized in the construction process must also be incorporated. In this context then, ethnodramas should be seen as *public voice* researches. That is, they are research scripts specifically composed for public consumption utilizing the colloquialisms and jargon of informants.[11,27]

Recognizing the need to construct ethnodramas that are simultaneously valid, cathartic experiences and representative of our repositioning of health research within the social contexts of postmodern critical theory, we have utilized a variety of ethnographic voice and text construction genres. Hence, ethnodramas are inclusive of *postmodern ethnographic reporting,*[8] *persuasion,*[18] *deconstruction,*[13] *hyperreality,*[4] *postmodern feminism,*[40] and *performance.*[11]

We envisage that the synthesis of these approaches in ethnodrama will be contiguous with reflexive techniques ranging from Woolgar's "strong program" inversion of the conventional subject/object relationship of the world,[49] to Popperian "falsification,"[38] and Kuhnian *notions of scientific elitism,*[23] as well as Tyler's (1988) *ethical reflexivity of representation.*[47] These provide a number of perspectives both in relation to and opposition with a Millsian concept of "ethical scientific practices."[33]

As a relatively advanced discipline, ethnodrama occupies that space which Denzin has described as improvisational, critical theater.[11] We combine "natural script dialogues" with dramatized scenes "crafted from fieldnotes" where the performance's textual narrative may actually consist of a composite gathering and editing of data which are fragmentary, autobiographical, personal, and incomplete. Thus, when performed, the text of the script, as dramatized by composite characterization, authentically invokes emotional, critical and personal responses from the spectators and the performers. This accessibility can be understood as relating the script's quality of readability to its inherent performance qualities. Together, readability and performability facilitate enhanced participatory validation processes, making the research report comprehensible to participants, some of whom are from marginalized groups, such as the mentally ill.[24,28]

On occasions when it has been necessary to link plot, subplot, and narrative, we have constructed fictional representations for inclusion into ethnodrama scripts. Although such inclusions are not the verbatim accounts of our informants, they are composites of data collected and authenticated by those informants as being atypical of interactions between health care patients and/or their health care professionals. That is, these linking sequences based on informant accounts or anecdotes were democratically agreed upon by project members as being realistic and plausible representations.[25,26] Where consensus was not achieved, the scenes were not utilized in the script, and in the instance of one play, the entire project stalled as participants were unable to democratically agree as to the validity and ethics of all sections of the text. This was a highly emotional scenario for the students who had been rehearsing and workshopping dialogue and scenes, and for the health professionals behind the project. With only days before the performance was due, the right to perform was removed.

A key issue in gathering ethnodramatic data, then, is who gets to have their voice heard and why. This is exemplified in the following excerpt from fieldnotes of one of the researchers of the play *Busting: The Challenge of the Drought Spirit*.[29] We believe that this excerpt clearly illustrates how the research process operates and highlights the critical aspects of choice and selection in the crafting of an ethnodrama script. In these examples, the central issue is that of the emotional import that visually performed research reveals to audiences of health performances, particularly the need to provide disempowered health care informants with a forum to have their agendas shown in an emancipatory and positive form.

> I walk from brilliant sunlight through dark tinted glass doors into the gloom of the narrow entry hall, I am late. I stand in front of this security screen so that Angie, the duty nurse, can peep through and identify me. The screen, which guards the admissions nurses from the outside world, is smaller than I remember. To its left — the entry door, to its right — Geordy. Geordy: 40ish-scowling, slight, unshaven, matted hair, is dressed in a once pale-blue cotton shirt and soiled, torn shorts. He sits on a narrow bench parallel to the reception desk. A strong, stale odor of alcohol and sweat wafts over me. I notice that one of his bare feet is caked in dried blood. The bench he occupies is purposely narrow, to prevent people from flopping down and sleeping on it, so he is forced to sit upright with his legs stretched out in front of him for balance. He repeatedly attempts to light a cigarette, seemingly unaware that his orange plastic lighter has no flame. He is becoming agitated. He dashes his lighter to the ground.
>
> "Bitches won't let me in," he bellows, as Angie arrives at the screen.
>
> "And that attitude won't get you back in here any quicker," retorts Angie as she opens the double security doors to let me through.

Suddenly the screen vibrates loudly as Geordy bashes it with his head....

"Stop that, you silly bugger," Angie commands. Geordy walks away, pauses, and bare-toed, kicks the wall hard in anger. He stands stock still for a second then hops in agony as his already battered foot fails to impress the masonry. He is now clutching his head in one hand and his toes in the other. If he were Buster Keaton or Oliver Hardy I might laugh. Angie does. "Geordy, behave!" she orders, trying to stifle her giggles. Meekly he sits once more on the bench and nurses his head then his toes.

"We only discharged him this morning. Busted before you could blink. Cardinal rules 1 and 2: no admission for persons known to be violent or for persons who have been through detox within the previous 24 hours." She laughs again then quietly adds, "I'll let him in when the piss and bad manners are out of him."

—Fieldnotes, March 1993.

These fieldnotes from the *Busting* research project were not used in the final script although the information gleaned from the incident did extend the participants' overall understanding of the phenomenon being studied. The decision to exclude this material was based on the scripting team's and other informants' opinions that rather than providing an empowered alternative view into the plight of those suffering from alcoholism, the inclusion of such data would reinforce stereotypical perceptions of alcoholics.

Ethnodrama should be seen as a collective attempt to provide a public voice forum for those most commonly identified as belonging to the disempowered fringes of health care: drug and alcohol abuse, rape, and mental illness. In our quest to give voice to those located in the shadow areas of health care, it can thus be seen that we have employed Augusto Boal's[6] theories about informants seeking political solutions to the roots of their political and social oppression. Boal postulated that invaluable insight may be achieved if one focuses on informant objectivity combined with collective action/response to conditions of social injustice. His audiences acquired authority through their reconstruction and redefinition of the meaning of past examples of their oppression and disempowerment in Boal's forum theater. Here then, is a further example of the academic use of theater to therapeutically empower informants by allowing them to present an alternate view of historical, social, and cultural slurs.[42a]

The identification and evocation of emotional responses[14] through staged presentations is further intended to confront audiences[9a] and participants with concrete instances of informant-lived experience. The intent to engage audiences at an emotional level is supported by the audiences' understanding that the stage representations they are witnessing are also documented, informant-worded research narratives based upon real lives. These narratives are public explanations that have been negotiated and interpreted in order to seek change and educate.

To assist in this understanding, and unlike filmic accounts, the audiences of ethnodramas are provided with psychological leaps from location to location by (a) program notes demarcating scene settings, and (b) most typically, copies of the performance script. The object of the exercise is not naturalistic theater but to give informant explanations, seek new understandings by challenging existing ones, and to give audiences opportunities for discussing and forming consensus on the issues presented to them.

The major advantage and uniqueness of ethnodramatic reporting lies in its accessing of mainstream consciousness through the *collective construction* of uncovered themes or data within a performance script. Using symbolic interactionist principles of the *"looking glass self"* in the creation of drama provides authentic context and meaning, enabling us to incorporate and imbue, in a democratic process of validation, participants' perceptions of self and others in our performances.[7] This grants us access to a greater potential audience for the dissemination of qualitative research,[34] greater currency for the notions expressed, as well as the opportunity for social processes relating to informing, educating, transforming, and facilitating catharsis,[33] and for epiphanies.[10] Extending Bauman's[5] view of performance as a specialized form of communication which embodies and enacts the myriad ways in which we inform and interact with others, we suggest that in ethnodramatic performances, script writers and performers have the added obligation of authentically expressing verbatim informant accounts of their lived experiences to audiences.[9] Further, our research supports data which debunked widely held stereotypical preconceptions about substance abuse. For instance, we explored the traditional notion of monosubstance abuse in relation to questions of polysubstance abuse. Similarly, in the course of our research, we have uncovered certain hidden truths about the unseen world of women victims of alcoholism, including their lack of access to adequate (or indeed any) sheltered accommodation compared to their male counterparts who are, by comparison, well catered to.* What we are proposing here, then is that the ethnodramatic form should be understood as highly flexible, allowing a range of distinct research approaches to be represented in an integrated manner.

In this way, we can identify the utility of ethnodrama and indeed the boon that performance ethnography of any kind may offer academia. The use of performance modes is easily recognized as being of significance to studies in symbolic interactionism as it seeks to demystify the social context

---

* We refer to specific contexts within Queensland, although this understanding may have some universality.

in which specific symbols (for example, hypodermic syringes) acquire meaning in the lives of substance abusers and their attendant health care providers.

Within the context of our research the potential of the symbolic to have a significant impact upon health care patients was clearly spelled out for us when rehearsing for the project performance *Busting: The Challenge of the Drought Spirit*.[29] An actor portraying needle-related behaviors in a detoxification unit attempted to discard an apparently used needle within a "safety sharps box" while demonstrating the routine use of multivitamin shots in detox treatment. Unfortunately the box fell to the floor, spilling three needles onto the front of the stage. As authors and producers, we did not recognize the potential impact of this symbolic reference (that is, needle-fixation) to many intravenous drug users. Some members of our special validating audience of informant detoxees from a local halfway house stood up to better see the rolling needles and gasped at their own behavior when they recognized what they had done. Consequently several informants became restless and revealed their ensuing preoccupation with the needle scene and its importance for them. One informant described a total incapacity to process any further aspect of the performance from that incident onward, such was the strength of association that the image of the needle stirred. Another mentioned the profuse sweating and anticipation that was triggered. Although this scene may amply demonstrate the effectiveness of the accurate portrayal of experience it also clearly highlights the power of symbolic interaction, suggesting that one must question the impact of this validation in terms of the recuperative processes of our informants.

However, missing from the dynamic are the palpable influences that performance research reporting may produce for certain audiences. As researchers who utilize public performance ethnography to encourage audiences to develop attitudes of "critical reflexive action,"[31] we are aware of its potential for raising government and public awareness of the plight of those suffering from mental health problems, substance abuse, or as a consequence of rape. We are cognizant of the fact that participants' responses may provide very useful insights for those in control of the purse strings which fund public health care, yet we are forced to concede that on occasion "never was success so painful."[31] To elaborate, consider for a moment Artaud's early *theater of cruelty* — a theater which intentionally challenged the psychological and emotional positioning of audiences in order to bring about a cathartic form of therapy.[2] It is far from clear that his works produced what might be considered a constructive catharsis. Without doubt, Artaud's performance genre represented a desire to produce deep emotional turmoil[43] and cathartic responses from audiences,[2] but his writings implicitly represented opportunities for audience psychological and emotional devastation as well as creation.[19]

Until recently we have been unable to embrace the impact and life-determining influences of performances upon audiences. Recent experiences, however, have moved us to herein explore the ethical dimensions of constructing critical research-based theater and catharsis. We have become aware that in the collaborative process that informs the construction of ethnodramatic performance, the responses of all involved may invoke deeper/hidden, personal and self-conscious analysis. Thus, while we have sought to explore and articulate "our relations with/for/despite those who have been contained as Others"[16] (in our case, health informants), to empower them we may have also inadvertently caused those most vulnerable to be placed in a position from which they could perceive no safe exit.

Ethnographic research performance in symbolic interactionism has been extremely well described[11,15] and exemplified by a range of prominent authors such as Clough, Richardson, Ellis, Bochner and, Mienczakowski, and Morgan. While narrative modes remain, understandably, the more common drama, poetry, performance art, visual art, song, video, and rock-video formats have each emerged within the performance canon of representational modes and contribute accordingly.[11] Although identifying strongly with this emergent tradition of research representation, we posit that these advances must be considered in conjunction with recognition of new and attendant ethical difficulties that inevitably exist. As one of the key objectives of ethnodrama is to provide a public voice for the previously hidden and, one could argue, oppressed world of particular groups of health care informants, we consider it imperative that our approach to the collection and reporting of data be conducted in ways which are neither dishonest nor exploitative of those people.[39] Our concerns are in many ways similar to those expressed in the ongoing debate regarding the ethics of conducting research via the Internet, as described by Thomas,[45] whereby some of the current issues being debated may have been unimaginable to even a recent generation of researchers.

The intention of this chapter then is to utilize examples from the authors' direct experiences to illuminate areas of concern for performance-related qualitative inquiry. This is by no means intended as an inclusive treatment of the topic, rather as an indication of a small range of the concerns and problems that we have identified at this developmental stage. Thus the primary intention is to wave a cautionary flag indicating perilous waters ahead, while also attempting to construct cogent approaches toward solutions. Like Thomas and other symbolic interactionists, we find ourselves compelled by Toffler's (1980) notion of the "Third Wave"[46] and particularly his identification that the future arrives in unplanned ways — which is an apt way of understanding our predicament as we seek to retrospectively negotiate a safe performance mode for qualitative inquiry.

Ethics and truth

Clearly, even in the most successful performance one cannot guarantee that each member of the audience will be satisfied in a transformational or even entertaining way. We cannot guarantee that our actors (usually students) will not be impacted through emotional revelation or otherwise transformed in some way. But since we hold an encompassing framework of social criticism, seeking to impact favorably upon social circumstances for marginalized persons with whom we cite allegiance, then we must take seriously the notion that these events and thereby these projects may have caused some harm.

Aside from the practical concerns of completing the validation prior to a full performance phase, the ethical problem of the possible (re)action of audiences postperformance looms largest of all. It has been broadly inferred by all participants that recognizing the polarities of argument within performance may lead to personal doubts in the minds of affected informants, performers, or audiences in regard to current or potential courses of action. Thus, for victims of rape for instance, the performance may affect their decision to go to the police or not, and have consequences thereby upon recovery and healing. Since we cannot guarantee what the correct course should be based upon experience or inquiry, and fear the causation of "doubt" in a person, or of some hesitancy in action, then the script must be silenced at this point. We cannot countenance a work that may lead to the wrong choice for some women. Thus we see that ethnographic performance carries rather heavy responsibilities.

The utilitarian notion that the greater good may be served by illuminating issues for debate and discourse, as contrasted with the possible confusions that may arise in some audience victims, is neither supportable nor demonstrable. We can find no reference to the representation of women recovering from sexual assault within any of John Stuart Mill's formidable texts — perhaps evidence of his sagacity?

Validation

Ethnodrama was conceived as a vehicle for the theatrical presentation of publicly accessible health care research data to inform governments, business, and the general public through a unique qualitative research methodology. As an emancipatory medium presenting data which challenges commonly held perceptions of the social and symbolic interactions of health care informants from marginalized groups, a foundational proposition for acceptability of ethnodramatic research within the arena of health care has been the determined process of validation. That is, the continuous process of taking the report back to its contributor-participants at the levels of thematic uncovering, script, rehearsal, and performance phases, to ensure the accuracy of its "re-animation."[34] As previously reported, ethnodramatic scripts are contextual reports of the largely verbatim data provided by informants. Any amendments to these data are undertaken in a collaborative forum that seeks to continually reaffirm the validity of the information presented. This vigor has been perceived by all participants as a key to maintaining high levels of script and performance veracity and credibility. Yet herein lies the crux of the ethical difficulty we have occasionally encountered: validity does not equate with an ethical basis for representation or performance. An accurate portrayal is not necessarily an ethical one. Consequently, we believe that some people may be damaged by exposure to some drama or ethnodrama on some occasions and under some circumstances.

Impact of fictional and dramatic suicide

Our team-led evaluation report of the performance project *Tears in the Shadows* illustrates aspects of the emotive and ethical difficulty to which we have already alluded. Indeed our evaluation concluded that exposure to the performance of this project was the source of some likely harm or distress to a small number of audience members and performers (and possibly fatally). In addition, we found that it is possible to recognize the manner in which the topic of suicide was reconstrued by audiences in a new light, this possibly to the effect of contributing to imitative suicide or suicide contagion.

Suicide in Australia

Currently Australians are experiencing a heightened interest and emphasis within governmental health and welfare policy for the funding of projects which might be able to stem the tide of youth suicide. This emphasis has generated a great deal of intended suicide prevention activity across Australia, with a recent federal stocktaking by the Australian Institute of Family Studies[3] recognizing almost 1000 projects within the first National Youth Suicide Prevention Strategy. Aside from the notion that the prioritization of youth suicide prevention has encouraged a proliferation of programs with a dramatic focus, there seems to have been a similar aggregation of mainstream and alternative (fringe) theatrical groups who have adopted a suicide-oriented agenda and produced works with a suicide theme.

The problem, of course, is that there is little evidence in regard to the impact that such drama may be having — with the fear that such performance may be problematic in terms of impacting adversely upon their targeted audiences and indeed upon performers. This question underpins our consideration of ethnodrama and ethnographic performance research and can be addressed through an examination of this project evaluation.

Tears in the shadows

The impact of fictional and dramatic suicide

The theatrical work *Tears in the Shadows* was performed for a brief Brisbane season in late 1998. This piece was developed within what might be considered a broad ethnographic framework and sought to express the experience of living with a psychotic mental illness. It consisted of 12 discrete scenes representative of elements of psychotic experience as devised over a 6-week period by a relatively well-known theater group. Each scene was developed within the group and validated by the group, although deriving from the unique experiences of discrete individuals within this theater group, which primarily included people with serious psychotic mental illness.

The group that constructed the pieces also largely performed it, although with professional directorial and performance assistance. A producer, who was additionally an experienced mental health professional, also guided the group and performance. The intent of the piece was to entertain, express personal experience in an empowering and cathartic manner, and to generally illuminate the issues of living with a psychosis for general and specialist health audiences. While at face value this seemed to be a play of educational value about living with schizophrenia, the performance involved some surprising elements in regard to suicide which are of import.

Suicide and performance

The suicidal context began within the accompanying program, which noted the dedication of the performance to a young man, a previous collaborator with the group, who had recently committed suicide beneath an onrushing train. The opening of the performance then indicated the general theme of the work with a song, which included a lyric, sung as the cast assembled on stage. The lyric is now being retrospectively recognized as a suicide note.

> *Tears in the Night: by ....*
> *Its time to destroy the outsider*
> *Its time to move into my shell*
> *Its time, I shall show you my darkness*
> *Tears from my shadow, memory to come*
> *Its time, so much forgotten already*
> *Its time so much memory lost*
> *Its time, you will burn on this fire*
> *Tears from my shadow, memory to come*
>
> *Its time to destroy the outsider*
> *Destroy the outsider*
> *Destroy the outsider*
> *Destroy the outsider*
> *Destroy the outsider*
> *Destroy the outsider*
> *... the outsider*
> *... the outsider*
> *... the outsider*
> *... the outsider*
> *... the outsider*
> *The outsider is here!* (Cast points to selves)

—Excerpt by permission of the
Warren Street Theatre Group.

The performance lasted 52 minutes and explored the experience of psychosis in a bleak and tragic manner with scenes of rape, child abuse, and a graphic hanging suicide. The general bleak tone was notable across the performance, which markedly lacked scenes of hope. Although subjectively shocking, this was something of an aim of the group. In that regard, the evaluation of the audience was consistent, but while the audience did find it shocking, nonetheless they still overwhelmingly enjoyed it. Some even described the suicide scene as their favorite part.

Summary of outcomes

Of principal significance to our consideration of suicide and drama were the following findings culled from the overall report:

1. Some of the language and imagery of the script left the team uncertain of the intention of the piece. While it is difficult to clearly grasp the import of a piece solely from a script — especially when such a stylized approach is sought, it may also be the case that some aspects of performance may also have been misunderstood by audiences, especially in a piece with so many diverse elements and scenes. This may be particularly problematic in the presentation of suicidal material to unprepared audiences.

2. While it is difficult to offer comment on content to the authors — who must be viewed as the experts of their own experiences, the team felt that the overall tone of the script was rather bleak, as highlighted by the scenes depicting suicide and sexual assault.

3. Overall, the responses indicated that the project was received in a highly positive manner. The great majority of responses noted the new learning and insights gained — outcomes very much attuned to the initial aims of the project. In particular, this responded heavily to a greater understanding of the difficulty and tragedy of living with a mental illness.

4. The work had a rather deep and reflective impact upon at least a small proportion of the audiences, some of whom were led to *new personal or emotional connections* with the topic of the work.

5. *Not all of these people described positive emotional experiences with the work, a small but important number describing their distress or discomfort with scenes, aspects of the work, or its overall tone.*

6. Some of the more powerful scenes were most enjoyed. *The suicide scene was found to have been "most liked" by five respondents.* Positive responses were also received in relation to a stylized rape scene and a mother-child abuse scene, although the numbers indicating this were very low.

7. A large group of respondents felt that the work of the group could be expanded to more mainstream audiences, primarily through greater

marketing. Four respondents indicated that the work might be suitable for specialized audiences such as schools.

8. The overwhelming majority of respondents indicated that they would like the opportunity to discuss the work, post performance. (This was not formally available.)

Superficially, it was evident that as an educational or instructive work, communicating the experience of psychosis, the project exhibited efficient and effective elements, with respondents indicating new insights, knowledge, and experiences in regard to psychosis and the experience of having a mental illness. Yet at the same time, this must be contrasted against certain negative outcomes resulting from the bleak tone and unexpected elements that, though largely enjoyed, did meet with some distress in some number. The audience responses attesting to the impact upon them in terms of new insights, challenged conceptions, and emotional or cathartic changes brought through the work must be considered against the possible impact on the small number of respondents who indicated unease or surprise with this personalized outcome.

Unfortunately, there were other unexpected outcomes with this project in terms of possible imitative suicides associated with performance. These were neither anticipated nor planned for in terms of evaluation, but require comment in conjunction with our understanding of ethnodrama and performance modes.

In summary, (1) the play was dedicated to a group associate who had recently committed suicide, (2) it included a song that may have indicated a suicidal intent, (3) it contained a graphic hanging suicide, and (4) it concluded with a final-night postperformance hanging suicide of one of the writers. This was followed within one week by the hanging suicide of an associate of the group, although we are not able to verify attendance. Although tautologous, the difficulty in validating suicide-related data is a significant issue that affects one's ability to make assertive causal statements, as noted in detail by Schmidtke and Haffner.[42]

In this way, it must be noted there is no way of ascertaining the role that the dramatic performance played in any of the actual suicides, especially given the presence of other likely indicators to increase suicidal risk, such as schizophrenia. The point is simply that the potency of dramatic representation may be a contributing factor and that the representation of, or direct referring to, suicide within dramatic performance,[37] particularly for presentation in educational forums,[21,44] is worthy of particular reservation. If suicide is a topic worthy of consideration prior to representation, then what other topics need to be similarly considered?

While in no way wishing to propose a censorship upon performance, at the same time one cannot avoid a demand for enhanced responsibility in performance. Performance is an emotional environment in which catharsis is always possible. In this regard, a notion of guidelines to aid dramatists in the suicide area have already been proposed[37] and utilized within Australian contexts, yet these, too, perhaps overstate the case, responding more to their inevitability than any desire to encourage such performances.

Conclusion

Within the cases discussed, it is clear that there have been significant, negative emotional effects upon some audiences and performers adjoined with the possibility of adversely affecting the social actions of people exposed to the performance. In one case associated with *Tears in the Shadows*, it is possible that the performance at least contributed to a suicide fatality. This chapter is emphatic in the demand for ethical responsibility in presentation and that the rights of researchers and performers cannot usurp the rights of audience members and whomever may be impacted indirectly by effects related to the performance. This is amply demonstrated by the recognition that while rape may be discovered within a research project as being traumatic and horrifying, this does not mean that it is ethically acceptable or valid to demonstrate the fact on a stage, for example, to a school audience.

These representational concerns can be met with simple enough solutions, such as pre-warning audiences, screening audience members, debriefing, and for the presenters, an exercise in self-awareness to ascertain that a performance has not been too hastily engaged in, with later regret. This may even be formalized to some degree, although producing an intriguing situation of auto-ethnographers requesting their own permission to perform their own ethnographic research. However, such a process can do the ethnographer little harm, and further, some auto-ethnographic performance bordering upon the confessional may be considered ill-advised by some people. We all say things we regret and much less desire to perform them.

The impact of a performance upon an audience cannot be underestimated and must be anticipated by ethical ethnodramatists and other performance ethnographers. The adherence to research ethics is usually well grasped, but fails to then account for what may become the greater ethical dimensions of "performance" in performance work, and which require a second and perhaps deeper ethical examination.

Our own experiences of working in this performance mode have seen us begin to deeply question the emotional dilemmas we sometimes create for ourselves and others. Unlike other versions of research reports, dramatic presentation can be as emotionally loaded for audiences and participants as any gun.

References

1. Agger, B., *Critical Social Theories*, Westview Press, Boulder, CO, 1998.
2. Artaud, A., *The Theatre and Its Double*, Signature Ser. 4, Calder & Boyars, (V. Corti, transl.), London, 1970.
3. Australian Institute of Family Studies, *Youth Suicide Prevention Programs and Activities*, A.I.F.S., Melbourne, 1998.
4. Baudrillard, J., *Simulations*, Semiotexte, (Foss, P. et al., transl.), New York, 1983.

5. Bauman, R., *Story, Performance and Event: Contextual Studies of Oral Narrative,* Cambridge University Press, Cambridge, U.K., 1986, 3.

6. Boal, A., *Theatre of the Oppressed,* Theatre Communications Group, New York, 1985. (McBride, C.A. and McBride, M.L., transl.)

7. Blumer, H., The methodological position of symbolic interactionism, in *The Process of Schooling,* Hammersley, M. and Woods, P., Eds., Routledge, London, 1976.

8. Clifford, J. and Marcus, G., Eds., *Writing Culture: The Poetics and Politics of Ethnography,* Harvard University Press, Cambridge, MA, 1986.

9. Coffey, A. and Atkinson, P., *Making Sense of Qualitative Data: Complementary Research Strategies,* Sage, Thousand Oaks, CA, 1996.

9a. Conquergood, D., Ethnography, rhetoric and performance, *Quarterly Journal of Speech,* 78, pp. 80–123 and 342–343, 1992.

10. Denzin, N., *Interpretive Interactionism,* Sage, Newbury Park, CA, 1989.

11. Denzin, N. K., *Interpretive Ethnography: Ethnographic Practices for the 21st Century,* Sage, Thousand Oaks, CA, 1997.

12. Denzin, N. K. and Lincoln, Y. S., Eds., *Handbook of Qualitative Research,* Sage, Thousand Oaks, CA, 1994.

13. Derrida, J., *Of Grammatology,* John Hopkins University Press, Baltimore, 1976. (Spivak, G. C., transl.)

14. Ellis, C. and Bochner, A., Telling and performing personal stories: The constraints of choice in abortion, in *Investigating Subjectivity: Research on Lived Experience,* Ellis, C. and Flaherty, M. G., Eds., Sage, Newbury Park, CA, 1992, 79–101.

15. Ellis, C. and Bochner, A., Talking over ethnography, in *Composing Ethnography: Alternative Forms of Qualitative Writing,* Ellis, C. and Bochner, A., Eds., AltaMira, Walnut Creek, CA, 1996, 13–45.

16. Fine, M., Working the hyphens: Reinventing self and other in qualitative research, in *Handbook of Qualitative Research,* Denzin, N. K. and Lincoln, Y. S., Eds., Sage, Thousand Oaks, CA, 1994, 70–82.

17. Fo, D., *Accidental Death of an Anarchist,* 1987 adaptation by Gavin Richards from a transl. by Gillian Hanna, Intro. by Stuart Hood, Methuen's Modern Plays, Methuen, London, 1987.

18. Geertz, C., *Words and Lives,* Harvard University Press, Cambridge, MA, 1988.

19. Grotowski, J., *Towards a Poor Theatre,* Simon and Schuster, New York, 1986.

20. Jennings, S., *Dramatherapy with Families, Groups and Individuals: Waiting in the Wings,* London, Methuen, 1993.

21. Kalafat, J. and Elias, M., Suicide in an Educational Context, *Suicide Life Threatening Behav.,* 25, 123–133, 1995.

22. Konradi, A., I don't have to be afraid of you: Rape survivors in court, *Symbolic Interaction,* 22, 44–77, 1999.

23. Kuhn, T., *The Structure of Scientific Revolution,* University of Chicago Press, Chicago, 1972.

24. Mienczakowski, J., *Syncing Out Loud: A Journey into Illness,* 2nd ed., annotated, Griffith University Reprographics, Brisbane, 1992.

25. Mienczakowski, J., Reading and writing research, National Association of Drama Education NADIE. (NJ), *Int. Res.,* 18(2), 45–54, 1994.

26. Mienczakowski, J., The theatre of ethnography: The reconstruction of ethnography into theatre with emancipatory potential, *Qualitative Inquiry,* 1, 360–375, 1995.

27. Mienczakowski, J., An ethnographic act, in *Composing Ethnography: Alternative Forms of Writing,* Ellis, C. and Bochner, A., Eds., AltaMira, Thousand Oaks, CA, 1996, 244–264.

28. Mienczakowski, J., Reaching wide audiences: reflexive research and performance, *NADIE J. (NJ)*, 22, 75–82, 1998.

29. Mienczakowski, J. and Morgan, S., *Busting: The Challenge of the Drought Spirit*, Griffith University Reprographics, Brisbane, 1993.

30. Mienczakowski, J. and Morgan, S., Finding closure and moving on, *Drama*, 5, 22–29, 1998.

31. Mienczakowski, J. and Morgan, S., Stop! In the name of love and baddies, grubs and the nitty-gritty. Paper presented at the Society for the Study of Symbolic Interaction: Couch Stone Symp., University of Houston, Houston, TX, 1998.

32. Mienczakowski, J., Morgan, S., and Rolfe, A., Ethnography or drama? An account of a research performance project into schizophrenia. *Aust. Natl. Assoc. Drama Educ. J.*, 17, 8–14, 1993.

33. Mienczakowski, J., Smith, R., and Sinclair, M., On the road to catharsis: A framework for theoretical change, *Qualitative Inquiry*, 2, 439–462, 1996.

34. Morgan, S. and Mienczakowski, J., *Re-animation of the Research Report: Shaping Nursing Theory and Practice*, Monogr. ser. No. 2, La Trobe University Press, Melbourne, 1993.

35. Morgan, S. and Mienczakowski, J., The application of critical ethnodrama to health settings, *MASK, Victorian Assoc. Drama Educ. J.*, 16, 15–19, 1994.

36. Morgan, S. and Mienczakowski, J., Stop, in the Name of Love!, unpublished script, 1998.

37. Morgan, S., Rolfe, A., and Mienczakowski, J., *Exploration! Intervention! Education! Health promotion!: a developmental set of guidelines for the presentation of dramatic performances in suicide prevention*, Mental Health Serv. Conf. Proc., Hobart, Oct. 1998.

38. Popper, K., *The Poverty of Historicism*, Routledge, London, 1960.

39. Punch, M., Politics and ethics in qualitative research, in *Handbook of Qualitative Research*, Denzin, N. K. and Lincoln, Y. S., Eds., Sage, Thousand Oaks, CA, 1994, 83–97.

40. Richardson, L., The consequences of poetic representation: Writing the other, rewriting the self, in *Investigating Subjectivity: Research on Lived Experience*, Ellis, C. and Flaherty, M. G., Eds., Sage, Newbury Park, CA, 1992, 125–137.

41. Saldana, J., Ethical issues in an ethnographic performance text: The dramatic impact of juicy stuff, *RIDE (U.K.)*, 3, 74–100, 1998.

42. Schmidtke, A. and Haffner, H., Public attitudes toward and effects of the mass media on suicidal and self-harm behavior, in *Suicide and Its Prevention*, Diekstra et al., Eds., Lieden, New York, 1989.

42a. Schutzman, M. and Cohen-Cruz, J., *Playing Boal: Theater, Therapy, Activism*, New York: Routledge, 1994.

43. Sontag, S., Introduction, in *Antonin Artaud: Selected Writings*, Farrar, Strauss & Giroux, 1976. (Weaver, H., transl.)

44. Taylor, B., *Educating for Life: Guidelines for Effective Suicide Prevention Programs in Secondary Schools*, Taylor Education, Melbourne, 1998.

45. Thomas, J., Introduction: A debate about the ethics of fair practices of collecting data in cyberspace, *Information Society*, 12, 101–117, 1996.

46. Toffler, A., *The Third Wave*, Bantam, Sydney, 1980.

47. Tyler, S., *The Unspeakable: Discourse, Dialogue, and Rhetoric in the Postmodern World*, University of Wisconsin Press, Madison, 1998.

48. Wilbur, K., *Eye to Eye: The Quest for the New Paradigm*, 2nd ed., Shambhala, New York, 1998.

49. Woolgar, S., *Science: The Very Idea*, Ellis Horwood, London, 1988.

chapter ten

An act of subversion: night workers on the fringe of dawn — from bow-wave to deluge

Jim Mienczakowski, Stephen Morgan, and Lynn Smith

> I walk from brilliant sunlight through dark tinted glass
> doors into the gloom of the narrow entry hall, I am late.
>
> —Fieldnotes from an urban detox unit,
> Queensland, March 1993.

Introduction

In our experimentation with an innovative ethnographic form, through which we have sought to challenge our readers and audiences using our research-based offerings as the basis for ethnodramatic performance interactions, we may at times have had little understanding of the full range of implications of our creativity. Professional therapists have compulsory sets of guidelines and ethical protocols to which they refer, and may draw upon the wisdom and advice of peers and mentors by thumbing through the pages of their professional bibles. In the developmental and emergent spaces of social science research, which performance ethnography and health promotion — that which we have labeled *ethnodrama* — occupies, we are writing the guidelines as we go. That is, we are sometimes engaged in the retrospective application of efficacy to our research products: those scripts and performances which constitute the ethnodramatic report. This then, is not a confessional tale, but a recognition of some undesirable outcomes and new responsibilities emanating from performance ethnography.

At times, recently, we have been troubled with the changes in our selves. "Troubled" is too mild a term, *disturbed* would be more accurate. As teachers and researchers, the authors have shared much common ground. For example, we have always seen ourselves as night workers on the fringe of dawn. Night workers who under the strong light of day challenge, contest, and if it is not too strong or ironic a construction, *worship iconoclasm*. At this point, we must add to the paradoxical nature of our theme in order to seek clarity: to be a night worker you don't have to actually work at night. Lest this is becoming totally confusing, as life and research have recently become for us, we will try to explain.

Night workers is a term we have been using to embrace those occupations, professions, and activities which are both acknowledged as essential to social existence but are marginalized in the everyday experiences, thoughts, and recognitions of most people. There are and have been numerous ways of sociologically describing such activities, but we have equated our experiences with those of night-shift employment specifically related to our professional *savoir faire* and research habitus. In hospitals, night workers may be the folk who run the morgue or clean up the detritus of failing or broken human bodies. In schools, they are the teachers who work with the severely disaffected and disruptive or those heart-wrenchingly disadvantaged by the circumstances of their health. They are the ambulance drivers, the hospice nurses, and traffic police who daily witness human suffering; psychiatric nurses in residential settings and prison psychiatric wards who, mob-handed,* grapple with and sedate patients who are violent or hell-bent upon self harm; those who volunteer time to work in soup kitchens or help street kids, operate needle exchange programs, and safe houses for victims of family violence and abuse. The night-worker list is endless and we can all, no doubt, add to it.

* "Mob-handed" refers to those situations when a number of people, in this case health care workers, collectively have to "physically grapple with" individuals or patients to subdue them.

Within almost every stable community there seem to exist networks of caregivers and health professionals who act as an ostensibly invisible network between the worlds of the well and the unwell and those who cope adequately with the world and those who cannot. Beels[2] has referred to community support networks as invisible villages that act as reference groups for truly marginalized communities. Though Beels refers to a form of hidden community which acts as a kind of agency in preparing the marginalized to reenter society, we refer to those who would seek to mend but often can't. An East Indian colleague told us that, in effect, what we were describing was a caste — in our case a collective of untouchables. We hope that is not so. And yet, the evidence is there, as the following excerpt from *Busting: The Challenge of the Drought Spirit*[8] exemplifies, in the words of helplessness and despair, that implied sense of belonging to that covert membership of the caste of untouchables voiced by nurses working in the arena of substance abuse:

> *Max: [Pause.] Sometimes. Sometimes I feel alright about it. It's just — you know — the pointlessness of it all. Sometimes, you know, the lack of respect and job satisfaction, I … I'm thinking of going back into the psychiatric wards for a while — or just doing something else. Something different.*

> *Lisa: I've had other jobs Max. I used to work at a cake factory. Four years of it — part time — no responsibilities — and I loved it.*

> *Max: Truth?*

> *Lisa: But I'm always pulled back into nursing. I fell into alcohol and drug rehab when I couldn't stick the acute admissions in psych nursing anymore. Lured into it by the glamor, the prestige. But really it's the same as anywhere else. Ain't that the truth. The people that come here are actively seeking help and are prepared to say, "Look, I really need help and you know what to do," etc. Dealing with acutely psychotic schizophrenics who are terrified by whatever they are hearing or seeing can be more frightening — because you're more of a threat to them. That's the big difference for me. It isn't intimidating here because they've chosen to be here, and I like most of the regulars too. (Draws heavily on her cigarette.) If you don't watch out you'll be in here yourself and no mistake. You say you can't cope — but who can? (Turns back to Max.)*

> *Max: I stopped telling people that I work here.*

Lisa: Really?

Max: Cause they always get embarrassed or start talking about how much they drink; or about a relative they've got who has a drink problem. No one wants to know about what I do. It's just shit — they just want to check that they're not heading down the same path. "Oh I don't drink before lunch so I'm O.K." or "My husband likes to drink, just a couple of casks a day, but he can handle it. He's not an f'ing dero — he's got a good job...." But more than that I can't hack — that the problem just keeps on growing — it'll never get solved.

Lisa: It's not the sort of problem you can solve, now is it?

Max: Yeah, I know that. But you know what I mean. Seeing the same faces come back time and time again: knowing that the people we treat will come back again and again and again and again.

Lisa: That's not the way to look at it, Max. If they didn't come here they might not have any chance at all. We keep them alive long enough to give them another chance to find the right way to clean themselves up. Some of them make it — if they want to badly enough and if they can get the right support. Think about most of our counselors. Nearly all of them are recovering alcoholics who have been dry for years and years. You can't ignore the possibility of hope, Max.

Max: Yes I can. Harry just phoned.

—Dialogue between two nurses, Scene 2.

Resorting to another, more macabre example, we promise we will then make a point. A piece of embryonic ethnographic research one of the authors conducted in the mid-1980s involved talking to a number of medical practitioners about their training experiences. The anecdote we are about to relate is poignant for several reasons (not the least being that the informant was a raconteur of note and was later destined to become father-in-law to one of the authors). The example goes thus: 60 or so years ago, before surgical solutions to large aortic aneurysms (swollen arteries which inflate with blood, balloon, then burst) were routinely in place, we are reliably informed that it was common nursing practice to place a large tin bath under a terminal patient's bed. In this way, the expiring patient could encounter his or her final rupture secure in the knowledge that when they departed it would not result in the endless mopping of the ward's floors and disturbance to others.

Our response to this anecdote has always been informed by our personal identification with night workers. Hence it has not been one in which we properly address concerns of how patients came to terms with enduring their final days in the knowledge that the splash pan was waiting for them — although this would seem to be a proper and sensitive response. Nor have we dwelt long on determining whether or not patients *pictured* the pan filling, or panicked at the slightest sounds of rain suddenly tapping on the window-panes, or of dripping taps.

Apparently, when such diagnoses were played out, it was often the case that the swollen artery seeped and leaked rather than burst explosively. Who could tell what it might do until it actually did it? However, in all of the imaginings and narratives concerning death, illness, and dying, we doubt that any of us has ever prefigured our demise as involving total fluid drainage, noisily, into a tin bath! No. Quite simply, we wonder about the night workers. Who are going to slide the deceased into the pan and carry him off to be cleaned and laid out? How would this experience correlate with their overall experiences of their working and social lives? And as night workers of sorts, we have always felt kindredness with those who share such hidden, authentic, and unexplored experiences.

Credentialing ourselves

In our professional researching and teaching lives, the range of understandings generated through working with severely disruptive and disadvantaged children, in detoxification units and psychiatric prisons, in needle exchanges, children's courts, abattoirs, and departments of social services, has given us an iconoclastic approach to the values of the daylight world and daylight workers. We have identified ourselves as sporadically existing on the fringe of daylight: trying to explain the experiences of recovering from sexual assault or of alcoholism, mental health consumption, and even suicide-related behaviors to those who work in other realities but who may influence the experiences and working conditions of these hidden social actors.

When compared to the lives and experiences of some of the folk with whom we have worked, on the whole, we realize our daylight lives are great. We remember the extremes and unacknowledged experiences and realities that some night folk discretely endure and we then question the rule-making, context-inhibitory practices of the academia and of legislative and statutory bodies, which we often view as exhibiting unnecessarily inhibitory — yea, bureaucratic and retentive behaviors! The academy often represents a passive, voyeuristic approach to explaining the lives and experiences of others. Anthropologically, we as academics seek ingress into the worlds of "the other" in order to merely report, critique experience, and then, without responsibility or conscience, move on.

So now, if you are still with us, you may come to see why we are feeling troubled. We have been calling ourselves night workers to avoid the self-recognition that we have become poachers. We have poached, drawing upon earlier professional experiences and sources, to inform our current academic professional status. Worse still, we are seemingly poachers now turned gamekeepers, and we have been forced to confront this recognition simply because our symbolic interactions and representations through ethnodramatic research have sometimes produced unanticipated and disturbing responses.

There are distinct differences between being the emptiers of the tin bath, colleagues of the bath emptiers, witnesses to emptying of the bath, or simply distant voyeurs of the entire process. We have been there (in a sense) but are no longer there. Does this matter?

Ethnodrama as catharsis

Performed pathogens

In many ways, the exposure to theater which seeks to redefine a person's relationship to a particular personal, health, or social topic may be loosely understood as entering the therapeutic realm, at least for some people and under some circumstances. This is not an unlikely proposition, given the significance of psychodrama and, more recently, drama therapy as legitimate therapeutic agents, but these modes of action are most directly understood from the participant's vantage. It seems that transformational possibilities can also exist through observation alone, although within the context of deliberately interactive and attention-grabbing drama, a participatory role of the audience may also be understood. Much work is currently being undertaken in this field, but our early research has clearly noted that research-based health performances attract specific health-interested audiences that contain many persons who might normally (or otherwise) be disinterested in theatrical performances. The therapeutic encounter constructed in an ethnodrama performance may, consequently, speak more directly to those informed by their own circumstances of health or professional interest therein, than with audience members seeking theater solely as entertainment or as a form of aesthetic appeasement.[10] There are reasonable grounds to explore the audience as existing as a momentary but uncohesive group, at least in respect of being learning groups.[3,13] Given this, we can see the performance as representing a sort of group intervention, which can then be understood within a number of frameworks.

However, by adding to the equation the performance of a particular health scenario or pathogen by a health consumer group experiencing the circumstances and revelations of a particular illness (from within a personal range of experiences of that illness), you are potentially constructing a highly emotive and challenging representation through a *performed pathogen*. Given the understanding that health theaters attract predominantly health-oriented audiences (other consumers, caregivers, health professionals and the like),[6] it is feasible that a performed pathogen may be received by an audience also identified by a pathogen. For example, a cast of persons with schizophrenia performs a piece about their own experiences of mental illness to an audience of co-sufferers and/or caregivers, thus potentially uniting performers and audience through a shared identification with that specific pathology.

Syncing the devil

We have recently revealed[9] how, inadvertently, we had placed a student actor who retained fundamentalist religious beliefs in a situation of personal vulnerability. Our cast included a student actor who believed that schizophrenia was the manifestation of the devil speaking through possessed persons. This student actor had happily played the part of a psychiatrist discussing schizophrenia to numerous and diverse audiences during the run of a play concerning schizophrenic illness. In a final performance, to an audience of psychiatric patients in a psychiatric institution, patients clambered on stage and confronted the actress as if she were *a real psychiatrist*. We, as researchers, saw this spontaneous performance as authentication of the experience being portrayed: before our eyes the play was being redrafted and vitally enhanced through active audience collaboration. The subjects of the research were simply adding data. While the director was disconcerted (as was one of the authors), we gleefully observed that the mental health specialists on the team roared with laughter and got high on the excitement of the impromptu performance. Indeed among the cast on stage, those who were also student nurses were in their element and excelled while their student actor colleagues were disconcerted and phased. A very different and somewhat threatening scenario was unfolding for the actor of fundamentalist persuasion: according to her world view she was being confronted by the personification of the devil. The scene was abruptly terminated as she, distraught and considerably unnerved, rushed off-stage and vanished into the night.

Busting: the silent terror of stereotypes

Immersed in our activities as night workers we have stumbled across the poignant murmurings of women victims of rape, domestic violence, alcohol and substance abuse, and the whispers of their caregivers. We record their collective voices crying out for public acknowledgment and recognition that all too often the consequences of their afflictions are swept aside or rendered invisible. In pursuit of our night work we have unveiled their silent and hidden pathogens which manifest in various forms including alcohol and antidepressant drug abuse, and given the afflicted and their health caregivers the opportunity to confront (albeit through theatrical performance) those who decide what is or is not included in public health policy and programs.

Excerpts from Busting: challenges of the drought spirit

(Lisa is a nurse on the night shift in an urban detox unit)

> Lisa: (Slow fade until the level of lighting has been reduced
> sufficiently to overlay the actors' faces with slow sequence
> of slides showing people in drinking situations — including
> the collage sequences from Act 1. Lisa checks the contents
> of the drugs trolley and writes up her observation notes as
> she talks.) We need people in this unit who can offer women
> support. The women in here are silent drinkers, just like all
> women with dependency problems must be silent. And there
> are so many of them out there. Thousands and thousands of
> women in every city of this country are raging Valium
> addicts or Serepax addicts or temazepam addicts — and they
> don't know where to go to get help. Most of them don't even
> know that they need it: because they are women and women
> aren't supposed to be seen "out of control" or using some-
> thing to "get by"... And though we might get a few women
> through detox, not many can stay to get rehab and the kind
> of help they really need — because they've got kids and other
> responsibilities to deal with apart from themselves. So they
> come in more and more often just to clean up and then they
> slowly drift into chronic alcoholism. Just because there isn't
> anyone else to care for the kids and nowhere for them to go
> with their kids where they can get the help that they need —
> until the social welfare takes the kids off them.

—Busting, Act II: Scene 3, p. 76.

If we refer to another excerpt from Busting involving a trio of incorrigible male patients suffering from alcohol dependency we are able to highlight a common perception of women victims of alcohol abuse which is not usually applied equally to their male counterparts:
Act I, Sc. V The Green Bottles:

> Lisa: Just quieten down, please. You aren't the only
> one here, you know? (Enter Dora in wheel chair.) Look.
> (Points at Dora sitting quietly and contentedly. Exit Lisa.)
>
> Wayne: Aw, look. That's sad, a woman here. It's not
> right.
>
> Jason: Terrible.
>
> Glen: Oh, yeah.

Jason: Awful for her.

Wayne: Poor thing.

Glen: Yeah, it's a bit pathetic isn't it.

Jason: Shouldn't happen to women. Makes me mad it does, such a waste. Never see them in the pubs though?

Glen: Secret drinkers.

Jason: How could she let herself go like that?

Glen: No self respect. Very sad.

Wayne: Well, you know how some women can be, don't you?

(All nod.)

Glen: Sad, I reckon. She must feel very pathetic — reduced to being in that state. Embarrassing for her.

Jason: Awful. She looks bloody awful.

Wayne: Ill, I reckon.

Glen: Hopeless case.

(Throughout this Dora sits passively and contentedly. She smiles at the three men. They get dressed again.)

Compare the observations of the misguided voices of the male victims given above to that of a female contemporary — the informed voice articulating the lived experience of alcohol abuse from a woman victim's outlook: *Act II: Scene 1.*

Chrissie: Hello, my name is Chrissie and I'm an alcoholic. I thank the Lord that I haven't had a drink today.

I began drinking and taking drugs while at high school. It would be hard to say if my drinking had been more of a problem than taking drugs.

I have hurt my family and my friends and myself. I am actively apologizing to those whom I've hurt and am trying to put things right. I'm also, like Mark, a counselor at A.A. Youth.

For the last three years I have been sober and free of drugs and have quit a life of danger and shame which these addictions brought me to. Or rather, that my disease brought me to.

Inside this organization I have helped many with the disease of alcoholism — particularly girls. Though you don't see them in bars or public places there are many women who have been caught by the cunning and pervasive power of drink. (Pointedly.) I'm not a group leader like Mark — because I'm only a woman and women are given no place in the world of alcoholism, but I am going to lead women into disclosure today.

There are fewer places of accommodation for female alcoholics and fewer agencies for support. St. Vinnies for example, offers beds for men only, as do most of the voluntary establishments. Women who drink are not as publicly acceptable as men who drink and we are the hidden victims of alcohol.

I'm not going to preach to you but point out what we go through. Women, especially young women, are more vulnerable when coping with an addiction problem. Had I not turned to prostitution, to earn enough to feed my habit; to buy whacks and a place to sleep, I would have been dead long ago.

Early in my disease my family disowned me — they couldn't cope with my anger and dishonesty — and I was forced to turn to crime to support myself. Prostitution and robbery made up part of my life style and even so I couldn't support my addictions. As they say, "One drink is too many and a thousand is not enough." (Pause) I thank the Lord that this has changed because I have opened up my soul and examined my heart and come to terms with my past.

Today we have some new young members and it is to them I speak. Recognize what you are and learn from the examples given in these disclosures today. What you hear may be many different stories but they are all really the same story. The people who disclose to you are warning of their pasts which may become your futures unless you open up your hearts to change. Brother Mark. (Mark returns to the lectern.)

As an interesting aside, take Ernie's self confession of addictive behaviors that points to the poignant irreverence of self-understanding evident in many informant stories. Ernie's anecdote is a form of dramatic relief from the pain of the other self-revelations. Both are deep from within the hidden worlds of addiction and both are largely verbatim informants' accounts.

Busting: Act II: Scene 3, Chopped Livers

> Max: (Slides turn to images of the A.A sequence.) The old ones are sad too. We had an old guy of 83 in last year, at an A.A. session. He disclosed. He said, "My name is Ernie and I'm an alcoholic. I thank the Lord that I haven't had a drink today. I am happy to say that a dissolute life of sin is behind me. I haven't had a drink in 18 years, and I beat compulsive gambling 13 years ago. I ceased a life of compulsive shop lifting and crime 9 years ago and I have been an upstanding member of the church since then. Now if I could just quit the heroin everything would be fine."

Issues in suicide prevention: bringing out the tin baths

Of late, our fringe-dwelling night work has led us to review the works of others involved in portraying suicide prevention and mental health issues through performance. Besides proposing an unassailable representation of illness, who in the audience (apart from others who share experience of a pathology) could authoritatively question the representation of lived experience? Such performance groups recounting depression, illness, and traumatic expression may not necessarily be adept in utilizing the dramatic form, and more importantly, the audience may not be able to operate in the symbolic codes, languages, and idioms of the performance genre. Such interactions, as uncovered within our research and in our evaluations of the ethnodramatic performances of others[9] may lead to confusing and dangerous messages being transmitted through performances. The question we ask is: *to what extent should theater performed as therapy for its participants be allowed to impact upon nonexpert, young audience groups?* Where does the therapy of a performing group become an oppressive act toward an unprepared audience?

One of the fundamental tenets of suicide prevention and therapeutic work in general is to first "do no harm." Our analysis of certain drama in suicide work, particularly that purportedly targeted toward preventing youth suicide, is that this tenet may have been compromised, be it with good intention or not. The evidence, based on our research team's evaluation and analysis of various suicide prevention performances as commissioned by federal and state government agencies[11] is compelling enough through the actuality of a postperformance suicide in one case and through the confused, depressed, and certain self-deprecating responses evidenced in another.

Of course, there are inherent difficulties with this analysis. We can easily argue that the deceased in one case may have been actively suicidal previously, as we have no pretest. Indeed, his status as a recent survivor of suicide and his longstanding mental illness certainly strongly indicate a high risk of suicide. In the second case, we have no pretesting for suicidal ideation, impulsivity, self-harm, or anything else among the school audiences, so again the students may not have been affected at all. However given that we know that suicides often occur in clusters, and that discussing, and graphically depicting suicidal acts may prompt actual suicidal tendencies, it is not surprising that cast members and associates related to this production killed themselves by imitative hanging. Perhaps they should have renamed their play: *Hanging in the Wings?!#**

Our insistence on regulation is not founded in concerns for the adoption of a censorial approach as to what should or should not be performed for vulnerable audiences. Rather, we consider the mere risk of affecting so many young, impressionable students is enough to urge a cautious approach — which the performances evaluated did not! We further note the findings of a current Australian report into the use of suicide prevention performances in Australian schools[12] which essentially notes *don't do it — too risky.*

However we have looked at these issues, we have been unable to escape the proposition that it really makes no sense to offer dramatists *cart blanche* access to topics that may significantly influence the resulting conduct of some vulnerable groups. We find ourselves culpable having exposed through performance research suicide, depression, sexual assault, and mental health issues without full cognizance of what such work entailed for vulnerable health consumer audiences. It may be increasingly important for health dramatists (and the arts, in general) to assure rigor and credibility to an increasingly evidence-based and critical society.

Competing codes and audiences

Codification and its conflict with artistic sensibilities may lead to a bloody debate, but a necessary one. In the preceding chapter we argued in favor of codifications and proposed a set of guidelines for suicide performance as a starting point for increased awareness and developmental notions of responsibility in this arena. We hope that this chapter and the following diagrams may highlight our concerns and possibly one means to move forward to safer arts/health performance practices.

Concern 1

Recognition of the reflexive, reflective and cyclical nature of health theater [research-based] should be encouraged in order to acknowledge audiences as constructive and contributing to understanding. Audiences are not passive recipients of performed understanding. A theoretical location encouraging creation of insight, debate and changes in professional practice is required in order to deepen understanding of health based performance research. What costs are implied for health consumers, audiences and participants [the vulnerable subjects] in engaging in ethnodrama?

Figure A demonstrates the cyclical nature of the combined research, data collection, and validation process that we endorse and practice in ethnodramatic endeavors. The authors consider that this process provides health care

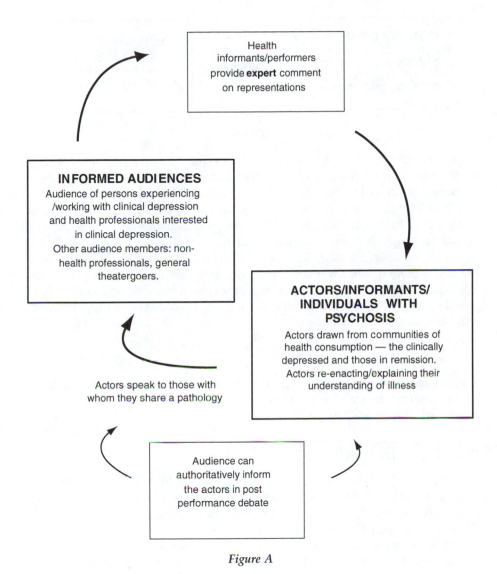

Figure A

informants with innumerable opportunities to authoritatively authenticate the script writing and performance construction process. This model enables us to actively encourage all participants (health care clients and practitioners, actors and audiences) of a performance to participate in postperformance evaluation and discussion *once* the initial script validation phases have been undertaken.

Concern 2

Conversely, Figure B models the one-way process which occurs when a group, ostensibly embodying authoritative understanding through lived

Figure B

experience of a specific pathology, seeks to perform its own personal therapy in front of nonexpert, student and other audiences. Such a limited process of consultation and validation leads us to question to what extent does the presence of clinical and social pathologies on stage inhibit postperformance discussion? In essence this is a didactic approach using research as a construction to push information onto general audiences. The plays hold research authenticity and, to a certain degree, are irreproachable. What redress is there to empower such audiences or to take care of at-risk audience members, that is, the unknown at-risk students, adults, or others not identified overtly as health consumers, the depressed students, or other of those with pre-diagnosed pathogens?

Concern 3

Should researchers and health performance constructors be responsible at all for what art does and what theater produces as emotional responses from voluntary audience groups !!??

Emptying the bath!

To demonstrate these concerns in practice we would now like to relate an experience of the authors in reviewing a performance. We were once called upon to critique a performance of *I'll Give You All the Diamonds in My Teeth.*[4] This was enacted by a university sophomore group and based upon the experiences of the author as a psychiatric nurse. Being a health-based piece it attracted a health-oriented audience that volubly protested outside the theater during the intermission. The audience, health professionals and patients from a schizophrenia support center, lived with, experienced, and understood the milieu of life with schizophrenia. What they witnessed was a performance dressed up as research which was stereotypical in its representation of illness, sensationalist, and clearly divisive. Members from this expert health group were affronted by the lack of recognition of their actual lived realities in a play which claimed a research basis. In effect, they felt that the performance would lead to further public misunderstanding of the realities of schizophrenic illness. They left the auditorium and staged a protest meeting, seeking to close the production down. What happened was typical of performances identified as research-based which are concerned with depicting a specific pathogen. The aims of the producers and author were incompatible with the professional experiences and knowledge of their health-based audiences.

Basically, the producers attempted to didactically inform an audience — which included persons with schizophrenia, their health caregivers, and health professionals — of how schizophrenia as an illness is experienced, claiming authority for their representation through research. The performance was heavily ironic and dramatized. The producers believed themselves to be competently presenting an understanding of the experiences of mental illness. Oops! Far from making a presentation to uninformed audiences (who may have accepted, uncontested, the suggestion that the staged presentation was research-based and therefore authoritative) the health-based performance attracted an *expert* audience who were able to contest the veracity of its representation.

Within the first 15 minutes it was evident that the producers had never been night workers and the bow wave of insight was rapidly becoming a deluge of offensive, oppressive, and stereotypical representations. After 20 minutes into the play, during an enactment of ECT (electroconvulsive therapy) treatment in which a seated patient screamed in agony as electricity buzzed through her body releasing her from psychosis and depression, an expert health consumer in the audience called out "Rubbish!" We subsequently learned that ECT treatment is not performed on conscious patients.

"Rubbish" as an audience catch phrase was soon supplemented with "fucking rubbish", "crap" and other less printable utterances. By the intermission a large group of health consumers in the audience raised a protest meeting in the carpark outside the theater. When they called "fetch the rope" we recognized that this performance had transgressed unwritten and previously unrecognized laws of unethical presentation. What is more, without a postperformance discussion element, audiences felt disempowered to correct or contribute to the understanding of what had been presented. Consequently, we find yet another lesson in the need for responsible construction and subject identification in written research-based performance genres.

Geertz's 1987 notion of *Being There* which is integral to the essential development of an audience's/reader's confidence in a text, promises the audience that the author has full insight and experience of the research field. This becomes more significant in ethnodramatic theater as the immediacy of the presentation wields enormous potential power over uninformed audiences through the interactive representation of pathogens. In effect, to describe a performance as research-based lends authority to representations that uninformed audiences may have difficulty in contesting. We urge our colleagues who are considering embarking on health-related dramatic performance to be wary, lest they too find themselves swamped by the deluge of waves of self delusion in the guise of professional insight.

References

1. Baume, P., Cantor, C., and McTaggart, P. G., *Suicides in Queensland: A Comprehensive Study*, Australian Institute of Suicide Research and Prevention, Brisbane, 1998.
2. Beels, C., *Survival Strategies for Public Psychiatry*, Jossey-Bass, San Francisco, 1989.
3. Johnson, D. and Johnson, F., *Joining Together: Group Therapy and Group Skills*, 6th ed., Allyn and Bacon, Boston, 1997.
4. Mazure, J., And I'll Give You All the Diamonds in My Teeth, unpublished play reviewed by John Edge, *The Bulletin*, 115(5883), 17 August, 1993, p. 91.
5. Mienczakowski, J., *Syncing Out Loud: A Journey into Illness*, Griffith University Reprographics, Brisbane, 1994.
6. Mienczakowski, J., The Theatre of Ethnography: The Reconstruction of Ethnography into Theatre with Emancipatory Potential, *Qualitative Inquiry*, 1(3), 360–375, 1995.
7. Mienczakowski, J., An Evening With the Devil: The Archaeology of Emotion, paper presented to the Society for the Study of Symbolic Interaction, August 11-12, Colony Hotel, Toronto, 1997.
8. Mienczakowski, J. and Morgan, S., *Busting: The Challenge of the Drought Spirit*, Griffith University Reprographics, Brisbane, 1993.
9. Mienczakowski, J. and Morgan, S., Finding closure and moving on: an examination of challenges presented to the constructors of research performances, *Drama*, 5, 22–29, 1998.
10. Mienczakowski, J., Smith, R., and Sinclair, M., On the road to catharsis: A theoretical framework for change, *Qualitative Inquiry*, 2(4), 439–462, 1996.
11. Morgan, S., Rolfe, A., and Mienczakowski, J., Exploration! Intervention! Education! Health Promotion!: A developmental set of guidelines for the presentation of dramatic performances in suicide prevention, in *Making History: Shaping the Future: The 1999 Mental Health Services Conference*, Robertson, S. et al., Eds., Standard Publishing House, Rozelle, NSW, 1999.
12. Taylor, B., *Educating for Life: Guidelines for Effective Suicide Prevention Programs in Secondary Schools*, Tailor-Made Education, Melbourne, 1998.
13. Yallom, I., *The Theory and Practice of Group Psychotherapy*, 4th ed., Basic Books, New York, 1995.

Index

A

Abuse, 21
  alcohol, 167, 187
  antidepressant drug, 185
  laxative, 93
  polysubstance, 168
  scene, mother-child, 174
  studies of, 11
  substance, 181
Academic status, 158
Addictive behavior, 188
Adolescence
  self-destructive, 99
  troubled, 88
After-the-fact reconstruction, 123
After-incident reports, 72
Alcohol
  abuse, 167, 187
  dependency, 27, 186
Alcoholics, recovering, 182
Analytic tools, emotions as, 129–143
  applying understanding, 141–142
  exploring emotions in qualitative
      research, 134–138
  lessons from psychotherapeutic practice,
      138–140
  social research and emotions, 132–134
  utilizing emotions within research
      process, 140–141
Anger, strategies for helping students cope
    with, 32
ANSOS, 48
Anthropology, 38
Anti-depressant, 94, 185
Anti-enlightenment approach, to experience,
    136
Archivists, 117
Attentive listening, 67
Audience, health-oriented, 192
Audiotapes, transcribers working with, 159
Autobiographical narratives, 65
Autoethnography, 85, 97
Autonomy, constraints on, 132

B

Bad feelings, 77
Bail hostels, 18, 22, 23
Balancing acts, regarding closeness to
    phenomenon under study, 9
Balinese Hinduism, 115
Belonging, sense of, 46
Biases, self-serving, 111
Bitter, Bitter Tears, 116
Blame, 31
Bulimia, 86, 88, 96
  conversation about lived experiences of,
    104
  history, 93
  intellectualizing about, 87
  peripheral to life story, 103
  private side of, 89
  rooted in perfectionism, 105
Burnout, 153

C

Camaraderie, 114
Cancer ward, ethnography in, 42
Caregivers, networks of, 181
Care-work, 54
Catch-22 situation, for women situated as
    academics, 133
Catharsis, 101
Cathartic changes, 175
Child, death of, 120
Childhood secrets, 92
Children's courts, 183
Circumspection, 101
Clinical encounters, personal reactions to, 56
Clinical interview, 64
Co-authored narrative, 6
Codification, 190
Collateral damage, 147–160
  alternative interpretation of process,
    153–154
  distal effects, 154